ENABLING
COMPREHENSIVE
SITUATIONAL
AWARENESS

Susan Lindell Radke
Russ Johnson
Jeff Baranyi

Esri Press
REDLANDS|CALIFORNIA

Cover: *Underlying map image by Esri with data from LOJIC, Louisville, Kentucky; upper photo from Shutterstock/Rob Byron; middle image courtesy of Emergency Communications for Southwest British Columbia Incorporated; bottom photo from Shutterstock/Digital Storm.*

Esri Press, 380 New York Street, Redlands, California 92373-8100
Copyright © 2013 Esri
All rights reserved. First edition 2013

Printed in the United States of America
17 16 15 14 13   1 2 3 4 5 6 7 8 9 10

Library of Congress Cataloging-in-Publication Data
Radke, Susan Lindell, 1957–Enabling comprehensive situational awareness/Susan Lindell Radke, Russ Johnson, and Jeff Baranyi.—
    First edition.
        pages cm
    Includes bibliographical references.
    ISBN 978-1-58948-306-4 (pbk.: alk. paper)
    1. Emergency management—Geographic information systems. 2. Emergency communication systems. I. Johnson, Russ, 1947–II. Baranyi, Jeff, 1975–III. Title.
    HV551.2.R33 2013
    363.34'80285—dc23   2013000976.

Ask for Esri Press titles at your local bookstore or order by calling 800-447-9778, or shop online at esri.com/esripress. Outside the United States, contact your local Esri distributor or shop online at eurospanbookstore.com/esri.

Esri Press titles are distributed to the trade by the following:

In North America:
Ingram Publisher Services
Toll-free telephone: 800-648-3104
Toll-free fax: 800-838-1149
E-mail: customerservice@ingrampublisherservices.com

In the United Kingdom, Europe, Middle East and Africa, Asia, and Australia:
Eurospan Group
3 Henrietta Street
London WC2E 8LU
United Kingdom
Telephone: 44(0) 1767 604972
Fax: 44(0) 1767 601640
E-mail: eurospan@turpin-distribution.com

Contents

Foreword

Emergency management means different things to different people. The government (federal, state, and local) is responsible for emergency management and public safety. Traditionally, the military is responsible for protecting the public from threats from foreign governments. The constant threat of terrorism has expanded and complicated the role of emergency management. This expanded role requires greater coordination among law enforcement, homeland security, and other first-responder organizations. It also requires the use of tools and technology, such as geographic information systems (GIS), to assist in complex problem solving, decision making, communication, and coordination.

Enabling Comprehensive Situational Awareness shows how GIS technology supports emergency management officials at all levels. It suggests how to configure and implement a GIS, taking into consideration technology trends, workflow complexities, and specific operations. The book also presents GIS and emergency management concepts.

GIS technology has evolved from a tool for complex analysis and general map production used by trained GIS professionals to a tool that can be used by nontechnical personnel. The first GIS applications were developed for workstations, later becoming desktop applications as personal computers became more powerful. GIS was somewhat "stove piped" for several years. With the advent of the Internet, computer software advances, and cloud computing, GIS has become easier to use, more accessible, and easier to customize, even for very specific workflow requirements. In many industries, GIS is now used routinely by operations-type personnel. Greater access to geospatial data and dynamic data feeds (cameras, weather, tracking, sensors, etc.) from multiple sources provides users with a current view of events. Within the homeland security, public safety, and emergency management domains, GIS is increasingly being used and expanded.

Enabling Comprehensive Situational Awareness examines how GIS has become a complete system and platform that consists of GIS tools (which align with the industry-standard National Incident Management System [NIMS], Incident Command System [ICS], and Emergency Support Functions [ESFs], which are identified and explained in the National Response Framework [NRF]), data management (including management of geographic data, text-based and tabular data, live feeds, communications, and other forms of data), and sophisticated display and analysis capabilities. This system enables deployment in the field, at the incident, and in emergency operations centers (EOCs). It connects government agencies and organizations. This combination of data, dynamic feeds, tool integration, and display capability helps emergency management personnel comprehend all aspects of an emergency situation and perform complicated analyses such as the following:

- How will a toxic plume spread and what will it affect?
- What intersections should be closed to control access into and out of an affected area?
- What infrastructure is affected?
- Where are the closest emergency support resources?
- What other incidents or events are occurring?

These capabilities integrate and extend many of the existing emergency management software systems used today.

Enabling Comprehensive Situational Awareness is the result of many years of lessons learned from actual emergencies, experiences in EOCs, and the ongoing and expanding needs of emergency management and personnel. It is a fresh look at an ever-evolving and powerful technology that is now being used by emergency management personnel to deal with complex emergency management challenges.

—Esri Public Safety Team

Preface

"Firefighters respond to incidents like this every day and they call us heroes for that. But now they are armed with information that GIS professionals, just like you, are providing to us and we come home safer because of that. In my book that makes all of you heroes just as much as anyone [of us]."

—Frisco Fire Department Assistant Chief Paul Siebert, July 12, 2010[1]

Enabling Comprehensive Situational Awareness addresses how geographic information systems (GIS) support the emergency management mission by bringing together geographic tools and data in a common operating environment. The ability to view vital information across agencies and between organizations enables more accurate decisions by emergency managers at critical times during the life cycle of a devastating event.

The emergency management workflow defines how response professionals plan for and react to significant events. The phases in this workflow include *mitigation*, *preparedness*, *response*, and *recovery*. GIS can be used in each of these phases through data management, planning and analysis, and field operations, thus allowing emergency managers to handle daily operations, complex emergencies, and disasters effectively.

While GIS resources have been available to emergency managers for quite some time, it is only now that the technology can be comprehensively implemented and integrated across all divisions within an emergency management agency and between event partners. Desktop GIS systems

---

1 Esri International User Conference, Plenary Session, San Diego, California, July 12, 2010.

have traditionally supported solutions to a range of disaster-related problems, but have historically required the skills of a highly trained GIS analyst to be integrated into the decision support system at an emergency operations center (EOC). Disseminating these solutions to managers and first responders making critical time-sensitive decisions has typically been an obstacle to the full use and integration of this technology in emergency management. Through the capability of the ArcGIS common operating platform, and adaptation of emergency management templates and applications available via the Local Government—Public Safety community on the ArcGIS Resources website, emergency managers now have direct access, via the web, to GIS data and tools that can enhance their awareness of an event as it unfolds in real time.

A web-enabled viewer within this common operating platform provides an environment where managers and duty officers are able to see, understand, and more effectively evaluate all of the geographic features that affect and are affected by a significant event. Customized, role-based live map viewers provide pathways into the data and out to the field, broadening information access across the field of operations vital to secure the safety of a community.

The Esri Public Safety team has worked over the past few years to break down, from beginning to end, the work that emergency management personnel do at emergency operations centers nationwide. In defining this workflow, they have paid particular attention to how and where GIS fits into the process to optimize operations. This work has been compiled into a rich collection of material focusing on the use and integration of GIS in emergency management. The team has presented this material at countless workshops, public safety summits, and training exercises across the country. This book brings these dynamic resources together into one comprehensive reference to further explore and share these fully integrated solutions with a broad audience of emergency managers, first responders, information specialists, and agency administrators. It provides the model of how the geographic approach (defined in this book) is best integrated into the emergency management workflow at all phases to enhance situational awareness. This information is supported by case studies highlighting the best practices and successes of various agencies across the industry.

The book is divided into six chapters, each providing a profile of the most effective use of GIS tools and analysis in emergency management. Chapter 1 introduces the organizational workflow and phases of emergency management planning and operations, including mitigation, preparedness, response, and recovery, and how GIS is used in each phase.

Chapter 2 focuses on how GIS manages the collection, storage, and movement of geographic data within and across an organization.

Topics include basemap data and imagery, operational layers, live data feeds, volunteer-generated information using popular social media portals, and planning and preparedness documents. This chapter also introduces a structured local government data model that includes features specific to emergency management and incident command.

Chapter 3 explores how planning and analysis using GIS enables effective emergency situational awareness. This chapter shows how, through the use of interactive analysis tools and results from more advanced spatial models, emergency managers can make decisions based on the results of risk and hazard assessment analysis; models of potential events and analysis of their possible consequences; and mitigation and preparedness planning, training programs, and budget planning.

Chapter 4 describes a vital component of the common operating platform: field support to and from first responders and recovery personnel. This chapter introduces the benefits of the shared working environment made possible by remote data collection via mobile GIS and crowd-sourced information and the use of mobile devices.

Chapter 5 draws the data, analysis, and field operations elements together into a set of role-based pathways through the GIS common operating platform. This chapter discusses best practices for configuring role-based applications based on user needs and workflows.

Chapter 6 focuses on implementing the pieces of the ArcGIS common operating platform that deliver information to managers and commanders. A selected set of templates and tools are explored that publish data embedded in the platform, bringing it to those making the critical decisions that save lives and protect property. Focus is on the implementation and configuration of key situational awareness viewers, maps, data, and resources. This essential information provides the emergency manager and support staff with what is needed to implement a system that will provide comprehensive situational awareness at all times.

Acknowledgments

Special thanks go to Rob Darts at the British Columbia E-Comm 9-1-1 and Steve Pallockov and Mike Brady of the Fire Department of New York GIS Unit for reviewing the manuscript particular to their work, as well as for hosting site visits and spending valuable time sharing firsthand experiences and on-site operations.

    Several people reviewed and edited selected sections of this manuscript, and others granted permission for the use of maps and graphics depicting best practices across the emergency management community. Thank you to Douglas Bausch at the Federal Emergency Management Agency (FEMA); Bruce Bishop at the City of Marietta, Georgia, GIS Department; David Blankinship at IntterraGroup and the Ventura County Fire Department; Tom Brockenbrough and Jason Loftus at the Accomack County, Virginia, Department of Planning; Renee Judkins and Ian Birnie of WorldView Solutions; Richard Butgereit at the Florida Department of Emergency Management; Dave Buckley at DTS; Curt Stripling and Linda Moon at Texas Fire Service; R. Vinu Chandran at the Cobb County, Georgia, GIS Department; Tara Byrnes Cordyack at GeoEye; Chris Emrich at the Hazards and Vulnerability Research Institute of the University of South Carolina; Jim Fox at the National Modeling and Analysis Center (NEMAC) at the University of North Carolina; Jessica Frye at the Kansas Adjutant General's Department; Robert Greenberg and Jeff Sopel at G&H International Services; Todd Gretton and Gareth Finney at Fire & Emergency Management, Victoria Department of Sustainability & Environment; Jeremie Comarmond of Esri Australia; Peter Hanna of the Baltimore City Fire Department; Jon Heintz of the Applied Technology Council; Matteo Luccio at Pale Blue

Dot Research, Writing, and Editing, LLC; Patrick Meier and Caleb Bell at Ushahidi; John Radke and Howard Foster at the UC Berkeley Resilient and Sustainable Infrastructure Networks (RESIN) project; Donna L. Roberts at the National Oceanic and Atmospheric Administration (NOAA); Buddy Rogers at the Kentucky Department of Emergency Management; Jean Schultz at the Iowa City Fire Department; Spencer Sweeting and Theodore Lemcke at IEM; Karyn Tareen at Geocove; and Rachel Tucker at the Food and Agriculture Organization of the United Nations (FAO).

We would also like to express our gratitude to Scott Oppmann and Lindsay Thomas of the Esri Local Government Resource Center team; Ryan Lanclos, Tom Patterson, Brenda Martinez, and Jesse Theodore of the Esri Public Safety Team; and the Esri Press team that supported the development of this manuscript.

Abbreviations

ADIOS2	Automated Data Inquiry for Oil Spills
ALOHA	Areal Locations of Hazardous Atmospheres
API	application program interface
AVL	automatic vehicle location
BGAN	Broadband Global Area Network
CAD	computer-aided dispatch
CAL EMA	California Emergency Management Agency
CAMEO	Computer-Aided Management of Emergency Operations
CATS	Consequences Assessment Tool Set
CDOG	Comprehensive Deepwater Oil and Gas Blowout model
CIMS	crisis information management system
CLU	Common Land Unit
COP	common operating picture
CRIMS	Critical Response Information Management System
CUSEC	Central United States Earthquake Consortium
DEM	Division of Emergency Management
DHS	Department of Homeland Security
DMA	Disaster Management Act
DTRA	Defense Threat Reduction Agency
DTS	Data Transfer Solutions
E2MV	Emergency Event Map Viewer
E2WS	Emergency Event Web Service
EADA	Esri ArcGIS for Desktop Associate
EOC	emergency operations center
EPA	Environmental Protection Agency
ERG	*Emergency Response Guide*

ERG2008	*Emergency Response Guidebook 2008*
ESF	Emergency Support Function
ESS	Emergency Services Sector
ETL	Extract, Transform, and Load
FARSITE	Fire Area Simulator
FDNY	Fire Department of New York
FEMA	Federal Emergency Management Agency
FSA	Farm Service Agency
GATOR	Geospatial Assessment Tool for Operations and Response
GeoCONOPS	Geospatial Concept of Operations
GIS	geographic information system
GISi	Geographic Information Services, Inc.
GMO	Geospatial Management Office
GNOME	General NOAA Oil Modeling Environment
GPS	Global Positioning System
GSTOP	GIS Standard Operating Procedures
hazmat	hazardous materials
HAZUS	Hazards–United States
HE	high explosive
HLS	Homeland Security
HPAC	Hazard Prediction and Assessment Capability
HSIP	Homeland Security Infrastructure Program
HTTP	Hypertext Transfer Protocol
HVRI	Hazards and Vulnerability Research Institute
IAP	incident action plan
ICP	incident command posts
ICS	Incident Command System
IHAT	Integrated Hazard Assessment Tool
IMSID	Incident Management Systems Integration Division
IRIS	Incident Resource Inventory System
JEM	Joint Effects Model
Kansas-MAP	Kansas Mapping Application Program
KDEM	Kansas Division of Emergency Management
LASER	Los Angeles Situation Awareness for Emergency Response
LGBM	Local Government Basemaps
LGIM	Local Government Information Model
MARPLOT	Mapping Applications for Response, Planning, and Local Operational Tasks
NAIP	National Agriculture Imagery Program
NBC	nuclear, biological, chemical
NEMAC	National Environmental Modeling and Analysis Center
NESC	National Exercise Simulation Center

NHSS	Natural Hazard Support System
NIMS	National Incident Management System
NLE	National Level Exercise
NMSZ	New Madrid Seismic Zone
NOAA	National Oceanic and Atmospheric Administration
NRCC	National Response Coordination Center
NRF	National Response Framework
PDA	personal digital assistant
PDC	Pacific Disaster Center
PSAP	public safety answer point
RESIN	Resilient and Sustainable Infrastructure Networks
REST	Representational State Transfer
RSS	Really Simple Syndication
SCC	Sector Coordinating Council
SCT	Secretariat of Communications and Transportation of Mexico
SEOC	State Emergency Operations Center
SE ROPP II	Southeast Regional Operations Platform Pilot, Phases I and II
SERT	State Emergency Response Team
SLTTGCC	State, Local, Tribal, and Territorial Government Coordinating Council
SMS	Short Message Service
SOAP	Simple Object Access Protocol
SOP	standard operating procedures
TIC	toxic industrial chemical
TIM	toxic industrial material
TxWRAP	Texas Wildfire Risk Assessment Portal
URL	uniform resource locator
USFS	United States Forest Service
USGS	US Geological Survey
VCFD	Ventura County Fire Department
VGI	volunteered geographic information
vUSA	Virtual USA

Chapter 1: Introduction

A tanker car carrying chlorine has overturned in Louisville, Kentucky. The dispatch officer on duty at the local emergency management agency logs this event into the crisis information management system (CIMS). The emergency operations manager on duty at the agency considers the gravity of the situation and decides to activate the emergency operations center (EOC) to evaluate and respond to the emergency. The manager is faced with a barrage of potential situations that will affect people and property across a wide area of the region. The first and most significant piece of information needed is exactly *where* this incident has occurred. Conventional 9-1-1 dispatch mapping applications can accurately locate such incidents on a street map. First responders can then be dispatched along designated routes to the site to secure the safety of people and property affected by the event.

A quick dispatch of first responders to the overturned tanker directly addresses the incident at hand, in this case, the chlorine spilling out of the tanker car. But what about the many other factors that influence how this event unfolds? Louisville is a large city in the Midwest along the Ohio River where flooding is a significant hazard. It borders the state of Indiana, so multijurisdictional issues are at play if and when a hazardous event occurs that crosses boundaries. Where are the roads and rail lines that transport hazardous materials? Who are the people that live in the neighborhood of the incident? Are there any schools, churches, or other locations nearby where large groups of people assemble? What is the traffic flow on adjacent roads at that time of day? Which way was the wind blowing when the incident occurred? These are all pieces of information that help the emergency operations manager characterize the incident and minimize the risk to the surrounding community. This information, when stored and managed as data that is tagged to locations on the earth (georeferenced) within a geographic information system (GIS), can be used to effectively support the entire emergency management workflow from beginning to end.

The 9-1-1 mapping application will route first responders to the site quickly, and they will do whatever is necessary to ensure public safety as soon as possible. But, knowing what is happening at the site of the incident and in the surrounding area is also critical to emergency response and incident management. Knowing the wind direction at the site of the tanker accident lets the emergency operations manager and the on-scene incident commander determine which neighborhood is affected by a potentially poisonous plume of toxic air. Knowing how many people live in that neighborhood ensures that the incident commander can determine fast and effective arrangements to relocate them to areas away from danger. Awareness of local traffic conditions around the site facilitates effective rerouting of passenger vehicles away from the hazard zone, while at the same time allowing emergency vehicles access to the site. This *situational awareness*, coupled with the expertise and experience that an emergency manager brings to the job, yields a professional who can make effective decisions in critical situations. The ability to make accurate judgments and draw rational inferences from situational knowledge at all stages of a significant event is what saves lives and secures property.

This knowledge, however, cannot secure the greatest public safety if it is not embedded in a well-defined organizational workflow that prescribes tried-and-true best practices for each emergency management phase of an event: *mitigation*, *preparedness*, *response*, and *recovery*. Each phase presents different parameters that challenge managers and operations personnel tasked with securing public safety. GIS offers tools designed to enhance situational awareness during each of these emergency management phases.

Emergency management workflow

Emergency managers are charged with protecting communities and reducing vulnerabilities. The consequences and costs of natural disasters, such as Hurricane Katrina and the Japanese earthquake and subsequent tsunami, and human-induced tragedies and catastrophes, such as September 11 and the Gulf oil spill, have taught us that we can no longer wait to respond until after a disaster occurs. The emphasis must be on preventing such events from occurring in the first place and reducing their consequences when they do. This mission has become increasingly complex because public expectations are high and resources and budgets are tight. GIS plays a pivotal role in bridging the gap between these limitations and the mission to protect public safety, as will be demonstrated throughout the rest of this chapter.

The emergency management workflow
Esri.

Mitigation is the first phase of the traditional emergency management workflow. It includes plans and actions to prevent hazards from developing into disasters, and to reduce their consequences when they do. In the case of the overturned tanker, mitigation measures include designating only certain routes away from population concentrations where tankers can travel during off-peak hours, thereby reducing the risk and extent of contamination if an accident should occur.

Preparedness is the development of action plans to enact when disasters and emergencies occur. If the tanker does overturn, posing a potential contamination hazard along a major transportation corridor, a well-defined preparedness plan would include the immediate use of effective tools to determine the direction and extent of the plume of toxic air, execution of evacuation plans, and designation of roadblock intersections.

Response is the ability to provide resources that help mobilize first responders and incident management personnel. When the tanker does overturn, how quickly a hazmat team can contain the potential contamination is determined by how fast personnel can get to the site and the effectiveness of their tools, resources, training, and information. A fast and efficient 9-1-1 dispatch system, coupled with situational awareness of current traffic and weather conditions and other potential hazards, shortens response time to the site. While police and firefighters work to secure the area and contain the immediate contamination, medical crews and specially trained community support personnel assist residents in relocating to safe assembly points where water, food, and medical supplies await them.

Recovery is perhaps the most difficult phase in the emergency management workflow. The actions taken to return a community to normal following an event, or, in some cases, better than before, pose a significant challenge to emergency management and operations personnel. Once the tanker contamination is contained and safety is returned to the community, steps are taken to clean up remaining debris, decontaminate and repair damaged infrastructure, and return residents to their homes. If, in the case of the tanker, long-term contamination conditions are possible, then arrangements must be made to secure the area for extensive recovery operations that may be disruptive to the community. On-site recovery operations are greatly assisted by mobile GIS field systems that monitor and record conditions on the ground in support of a full event recovery plan.

The challenges that emergency managers face in each of these phases of the workflow warrant smarter and better-designed solutions to secure public safety. Collecting large amounts of data to inform managers of existing conditions; performing complex risk and hazard analysis to minimize the chances that a disaster may occur, and identifying the vulnerabilities if it does; developing plans to prevent events, and lessen their impact should they occur; tracking and managing resources to expedite response; and disseminating information in and out of the field, as well as to officials, media, and the public as an event unfolds, are all challenges that require comprehensive situational awareness embedded in a geographic view of a community.

The geographic approach

The daunting list of challenges faced by emergency managers tasked with securing public safety can be overcome by a geographic approach to information management. Emergency managers need to access and collect extensive data from a variety of sources to make effective decisions throughout all phases of the emergency management workflow. In addition to basemap layers of the existing terrain and local infrastructure, data layers of traffic; live webcam video feeds; crowd-sourced information; weather; and police, fire, and medical resources and facilities are also necessary to effectively manage an emergency event. Using the geographic approach, the data can be collected to not only record *what* an entity is, but also *where* it is and what is nearby or around it. This "geo-enabled" information can be effectively integrated into and analyzed in a geographic database, turned into actionable intelligence, and then delivered through a *common operating platform* to stakeholders to be shared, viewed, communicated, and acted upon before, during, and after an event.

ArcGIS: The common operating platform Esri.

Common operating platform

A fully populated and integrated common operating platform forms the foundation of an EOC's intelligence. This powerful geo-enabled platform facilitates the deployment of a series of role-based *situational awareness viewers* that provide each agency and organization active in an event with a mission-specific live map of its own current situation embedded in the wider community. These viewers are engineered to support the power of GIS, seaming together a collection of basemaps, operational data, and analytical tools that align with each of the actively participating agencies and the Federal Emergency Management Agency's (FEMA) Emergency Support Functions (ESFs). Each agency and partner uses its specific viewer to ascertain the current conditions of its infrastructure and surrounding community to inform its recommended actions. These recommendations are then fed back to the central operations commander during briefing sessions. The compilation of all recommended actions, every one of them based on complete situational awareness of each ESF, comes together in a comprehensive *common operating picture* (COP) for the commanding officer who is charged with making the vital decisions that will secure the safety of the community.

In the case of the overturned tanker carrying chlorine in Louisville, Kentucky, comprehensive situational awareness is enabled when the many

The common operating platform architecture Esri.

previously unconnected datasets, workflows, and operations are integrated into one geographic platform providing a shared view of the region.

Within the central COP at the EOC, the geography of the Louisville area is delineated using basemap layers of the topography, streets, place-names, parcels, and aerial imagery. This gives the commanding officer a complete picture of all existing infrastructure in the region at the time of the event. Operational data layers are added that let the commanding officer quickly characterize the nature of the event and its situation with respect to current conditions and other events happening in the region. These data layers include geographic data services and feeds coming from the partner ESF agencies and organizations collaborating on the response. Current police and fire incidents, weather updates, social media, traffic incidents, live video webcams, planned events, and georeferenced news feeds that show the locations of current events in the national news services combine to provide a comprehensive picture of existing conditions. Each partner is equipped with a mission-specific map view of the event focused on the particular needs and requirements relevant to its emergency service function. Key analytic tools tailored to each function use the basemap data and operational information to geographically evaluate and determine the potential impact of the event and then present recommended actions to the central commanding officer, who can then quickly determine who is affected and what to do to minimize the risk.

In order for this common operating platform to be effectively used to support the emergency management workflow, a number of critical component patterns are needed, including data management practices, planning and analysis models, field operations and maneuvers, and situational awareness capabilities.

Data management

Emergency management personnel are required to collect data from agencies in the community and from the public to be as fully informed as possible. This inflow of disparate data is a huge challenge for managers who are responsible for assessing and characterizing vulnerabilities to protect the public and infrastructure across jurisdictions. The GIS platform can be used to organize data through its geographic location. Data is then available for analysis and operational use to support mitigation, preparation, response, and recovery.

Geo-enabled data contains many descriptive characteristics about features in an area of concern. In the case of the overturned tanker, these may include the names of streets under the plume of toxic air, and the names, addresses, and phone numbers of property owners living there. Accompanying this descriptive information is the coordinate position of each feature so it can be placed on a map in the correct location. The pairing of descriptive, or attribute, information with coordinate, or geographic, information transforms a database into a *geo*database, empowering an emergency manager with situational awareness of an event as it unfolds and affects the surrounding area.

Because all events occur at a location, GIS provides the common capability, as the lowest common denominator, to contextualize all of the data geographically. The greatest capability and security is realized when the data that characterizes a community, and an event that may be threatening its safety, is stored in an integrated geodatabase and managed within a GIS on a stable hardware and software platform that can accommodate the great extent of detailed databases and aerial imagery needed to be informed.

Data management does not only pertain to organizing attributes about the features on the ground. It also includes arranging and indexing documents and plans linked to locations in the field. Floor plans, documents detailing evacuation procedures, and photographs of buildings and conditions can also be managed within the GIS by their geographic relationship to features in the field. During stressful situations when an event is unfolding, fumbling around to find binders located in various offices across agencies wastes an emergency manager's time while the safety and security of the community is at stake.

A geodatabase has the capability to link together agencies and operations active in the event, such as incident command posts (ICPs) and support teams, regional and national agencies, and field personnel, including search-and-rescue operations.

One of the most difficult tasks in managing large emergencies is managing all of the resources needed to support successful incident management. The integrated GIS, empowered by a well-designed and maintained geodatabase, has the capability to access and manage such resources as medical response teams and emergency housing and food supplies that support large groups of people and materials as they converge on an affected area. Where are the closest (quickest to the scene) resources, what is their current status, and what is the most efficient way to move them from one priority to the next? GIS provides the ability to locate, manage, and assign the appropriate resources by presenting them logically, based on their location and proximity to the incident.

During significant emergencies, imagery also becomes critical to understanding the overall consequences of the event. Current high-resolution imagery of an area is used to detect changes in an area that may affect decisions regarding response and recovery. Imagery files are large and require a data infrastructure to catalog, process, and make them operational. A well-designed GIS provides the ability to organize, store, analyze, and manage imagery that can be accessed as required from the common operating platform.

GIS is an essential component in a good data management strategy. GIS provides the right data at the right place at the right time to support decision making in emergency operations circumstances. Access to data from other agencies via a system-wide (*enterprise*) GIS—be it from hospitals, hazards, utilities, transportation, the field, or live feeds—gives the greatest virtual, real-time picture of an event as it unfolds. Chapter 2 provides an in-depth review of the primary components that make up an effective data management model in emergency management and presents a range of solutions and best practices proven in the field.

Planning and analysis

Having all of the relevant data embedded in an integrated geodatabase is only the first piece of the puzzle. How to study and analyze the data to assess a community's vulnerabilities and how best to mitigate those vulnerabilities are the next pieces. GIS provides the analytical tools that allow scientists and decision makers to see how events and policies affect communities. Revealing land-use changes that occur in response to new transportation routes, tracking demographic shifts as economic conditions change in a community, illustrating the geologic profile of a region, and monitoring the flow of water within a watershed are only a few examples of how GIS is used to model the natural and human landscapes.

As typical as these methods of geographic analysis have been over the past two decades, they have for the most part remained within the domain of well-trained GIS professionals. Sophisticated models built using high-level programming languages ensured that such tools would not easily integrate into everyday decision support systems used by managers trained in professions other than GIS.

Through the use of a GIS situational awareness viewer, the emergency manager now has access to an embedded set of planning and analysis tools present in the Emergency Response Guide (ERG) toolkit. These tools enable the quick and effective analysis of georeferenced base data and operational layers to make informed decisions regarding who and what is immediately affected by an event. Advanced planning analysis and consequence models can also be leveraged by integrating the results of high-level spatial studies to assess where a community's greatest vulnerabilities

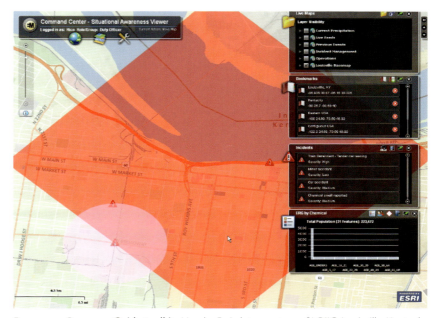

Emergency Response Guide toolkit Map by Esri; data courtesy of LOJIC, Louisville, Kentucky.

are, where mitigation priorities should be, and where contingency plans need to be developed. Through the use of a GIS situational awareness viewer, managers can also justify budget needs in support of mitigation plans by illustrating the potential consequences of an emergency event.

In the case of the overturned tanker, mitigation plans developed with the use of GIS can be put into operation effectively, allowing quick action to reduce the risk the overturned tanker will pose to the surrounding community. As emergency planners manage natural and technological hazards in the area, they can quickly deploy the embedded ERG to model the direction of the plume of toxic chemicals and calculate the number of people living within the affected area.

The situational awareness viewer also allows layering of natural and technological features, such as floodplains and critical infrastructure and locations of vulnerable populations at schools, nursing homes, and hospitals, on one map. Although this map may contain all of the relevant information needed to enable complete situational awareness, it can make for a complex map that is difficult to understand.

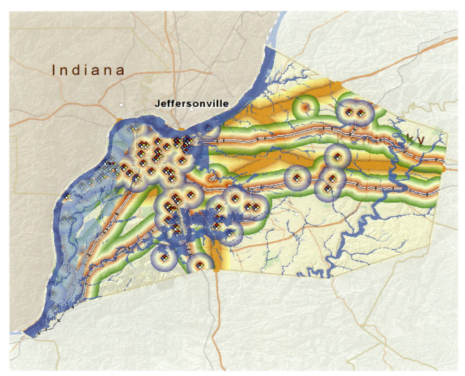

All natural and technological data layers for the city of Louisville, Kentucky, combined on a single complex map   Map by Esri; data courtesy of LOJIC, Louisville, Kentucky.

Analytical tools deployed in a desktop GIS let users move beyond the visualization of the data. Merging these maps together creates an understandable representation of complex situations and, in this case, displays the density of both natural and technological hazards that exist in the area. A weighted grid, generated on the desktop GIS using advanced spatial analysis tools, identifies concentrations of hazards, creating a high-level view that focuses on potentially vulnerable areas.

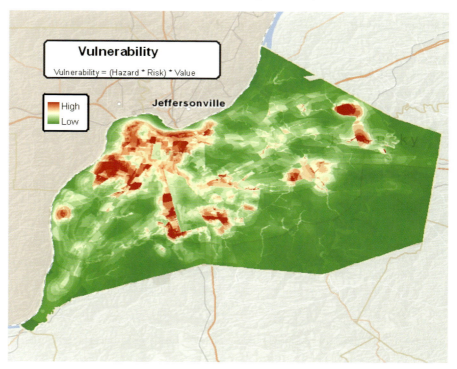

Combined vulnerability map of Louisville, Kentucky Map by Esri; data courtesy of LOJIC, Louisville, Kentucky.

The maps resulting from these analyses can be integrated into the common operating platform and made available to emergency operations personnel through their situational awareness viewer. Better and more-informed decisions can then be made based upon the potential threat to lives and property. Having the ability to integrate the results of risk and hazard assessments into the platform and to display them in the situational awareness viewer puts the good work of planners into an operational context. This improves upon the planning stages of emergency management, as well as upon mitigation and preparedness. The results of GIS analysis are now available and accessible beyond the domain of the highly trained GIS professional. They are now at the fingertips of personnel working at emergency operations centers and planning agencies nationwide.

These tools and analyses are fully explored in chapter 3. Vulnerability analysis is reviewed, and examples of how the results are used by managers in the command center and in the field are shown. Additional modeling examples presented include analysis of the potential effects of natural hazards, including earthquakes, floods, and wind events, and fire management strategies in urban and rural communities.

Field operations

Data management and planning and analysis make up the foundation of enhanced situational awareness. Once these critical pieces are in place, an EOC is then equipped to provide a functional common operating environment to emergency managers and operations personnel. But, emergency events don't occur in the EOC, they happen in the field. The managing officer's job is not only to coordinate the integrated response and recovery activities at the EOC, but also to support the incident management personnel coordinating the response effort at the site of the incident and evacuation areas.

Gathering data at the site of an event is the only way to truly gauge its effects and the expected resources needed for recovery. Field data collection has typically been marred by communication incompatibilities and inconsistent assessment tools. The use of mobile GIS—now available on GPS devices, tablets, and smartphones—allows field operations personnel to have access to the same information on handheld devices in real time that the event partners have access to at the EOC. This information is stored in the geodatabase embedded in the common operating platform used by all event partners in all places throughout the span of an event.

In the case of the overturned tanker, emergency response personnel in the field have delineated two divisions around the perimeter of the affected site, and an incident command post and staging area are established off-site in a nearby parking lot. Because the data management system already includes an integrated mobile GIS support capability, this information is automatically available to the managing officer at the EOC through field data synchronization to the common geodatabase. Now the managing officer can refer to the same information in the situational awareness viewer that the response teams in the field are using to quickly find resources, such as the closest mobile feeding kitchen, that incident management personnel need to effectively do their jobs.

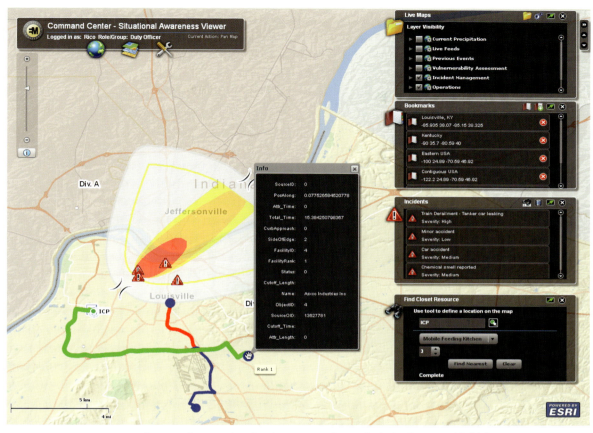

Command center situational awareness viewer displaying integrated field data

Map by Esri; data courtesy of LOJIC, Louisville, Kentucky.

Once the incident is under control, and the safety of the people in the area is secured, the managing officer assesses the extent of the damage caused by the spill. The recovery team is deployed to the site with handheld GPS-enabled mobile GIS devices to identify and record the extent of the spill and level of damage to each structure. They digitally tag structures in the vicinity and record remarks to characterize the damage to buildings and other infrastructure. Buildings are color-coded to indicate levels of damage and documented with photos to support the damage designation (see figures on pages 14-15).

In addition to capturing and visualizing these data in the EOC's situational awareness viewer, and in the field on handheld devices, this real-time data can be quickly prepared as digital map presentations or printed as presentation-sized paper maps to support coordination during briefing sessions.

This coordinated field work that is synchronized with the common EOC geodatabase is immediately available to the managing officer, who can now generate damage assessment reports based on work done by teams on site. The officer can quickly assess and summarize the financial effects of the event when requesting recovery funds from government agencies (see report on page 16).

The speed and ease with which a damage assessment report can be generated using a mobile GIS system integrated within a common operating platform sidesteps many of the typical bottlenecks associated with seeking a disaster declaration necessary to secure recovery funds, and ultimately to return a community to safety.

Chapter 4 reviews the many opportunities that mobile GIS offers in support of comprehensive situational awareness in emergency management. Examples include the use of mobile GIS in urban and wildfire response scenarios in Georgia, Texas, and California and recovery and damage assessment operations in the Australian bush fires, Tuscaloosa tornados, and Gulf oil spill.

Handheld GPS-enabled mobile GIS device displaying a common operating map

Map by Esri; data from US Geological Survey, device courtesy of Trimble.

Handheld GPS-enabled mobile GIS device displaying a damage assessment form

Device courtesy of Trimble.

Command center situational awareness viewer displaying damage assessment field operations

Map by Esri; data from LOJIC, Louisville, Kentucky.

Situational awareness

Once the data and analysis tools of an effective emergency management system have been designed and implemented, and the work of field personnel is remotely integrated into the common operating platform, a complete and comprehensive level of situational awareness is achieved. The benefits that this level of awareness has on a managing officer's decision-making capabilities cannot be overstated. To be able to consider all relevant variables in an emergency situation greatly increases the chances that the effects of a catastrophic event will be minimized. Let's review how this opportunity plays out in the case of the overturned tanker in Louisville, Kentucky.

Each agency of the City of Louisville maintains its database on a daily basis to manage the city. Together, these agencies manage basemap data of the geographic features of the area, as well as the operational layers of information regarding daily incidents and planned events around the city. As described earlier, when an event such as the overturned tanker occurs, the city's computer-aided dispatch

Confirmed Damaged Structures
3/11/2009

If your property has suffered damage due to the chemical release and does not appear on this list or the map, please call the Emergency Management office at 502-555-1212. Please do not reenter any structure until it has been deemed safe by the health department.

APN	Street Address	City	Zip	Type	Damage
014B00820000	1311 W MARKET ST	LOUISVILLE	40203	Residential	Moderate
014G00250000	1351 W MUHAMMAD ALI BLVD	LOUISVILLE	40203	Business	Light
014A00500000	1612 W MARKET ST	LOUISVILLE	40203	Residential	Light
014A01290000	1637 W MARKET ST	LOUISVILLE	40203	Residential	Moderate
014B01180000	1214 W MAIN ST	LOUISVILLE	40203	Residential	Heavy
014C02120000	1030 W MARKET ST	LOUISVILLE	40203	Residential	Heavy
014G01750000	1420 W JEFFERSON ST	LOUISVILLE	40203	Business	Moderate
015F01200000	1618 COLUMBIA ST	LOUISVILLE	40203	Residential	Moderate
015F00630000	1723 ROWAN ST	LOUISVILLE	40203	Residential	Light
015F00750000	1709 ROWAN ST	LOUISVILLE	40203	Business	Moderate
014F01420000	1518 CEDAR ST	LOUISVILLE	40203	Business	Heavy
014B00250000	1328 W MARKET ST	LOUISVILLE	40203	Residential	Light
015A00610000	1020 W MARKET ST	LOUISVILLE	40203	Residential	Light
015A00470000	1601 W MAIN ST	LOUISVILLE	40203	Business	Light
014A01260000	1704 W MAIN ST	LOUISVILLE	40203	Business	Moderate
014A01620000	1537 W MARKET ST	LOUISVILLE	40203	Business	Light
014B01010000	1205 W MARKET ST	LOUISVILLE	40203	Business	Light
014C01090000	1145 W MARKET ST	LOUISVILLE	40203	Residential	Moderate
014A00660000	1516 CONGRESS ST	LOUISVILLE	40203	Business	Moderate
014G00930000	1328 W LIBERTY ST	LOUISVILLE	40203	Residential	Heavy
014A01300000	1635 W MARKET ST	LOUISVILLE	40203	Residential	Light
015A00640000	1528 ROWAN ST	LOUISVILLE	40203	Business	Moderate
015F01260000	212 N 16TH ST	LOUISVILLE	40203	Business	Light
015A00810000	1721 CROP ST	LOUISVILLE	40203	Residential	Light
014A01270000	1700 W MAIN ST	LOUISVILLE	40203	Business	Moderate
014A01480000	113 S 17TH ST	LOUISVILLE	40203	Residential	Moderate
014A00820000	1505 CONGRESS ST	LOUISVILLE	40203	Residential	Heavy
014G00920000	1330 W LIBERTY ST	LOUISVILLE	40203	Business	Light
015F01110000	1607 ROWAN ST	LOUISVILLE	40203	Business	Moderate
015A00830000	1520 BANK ST	LOUISVILLE	40203	Residential	Light
014A00650000	1511 W JEFFERSON ST	LOUISVILLE	40203	Residential	Light
014F00560000	1626 CEDAR ST	LOUISVILLE	40203	Residential	Moderate
015B00740000	223 N 15TH ST	LOUISVILLE	40203	Residential	Moderate
015A00300000	1710 ROWAN ST	LOUISVILLE	40203	Residential	Heavy
015B00080000	1301 W MAIN ST	LOUISVILLE	40203	Residential	Light
014A01160000	114 S 17TH ST	LOUISVILLE	40203	Residential	Heavy
014A00810000	1509 CONGRESS ST	LOUISVILLE	40203	Business	Light
015F01250000	1608 COLUMBIA ST	LOUISVILLE	40203	Residential	Light

"Confirmed Damaged Structures" report needed to request recovery funds from government agencies Report by Esri; data from LOJIC, Louisville, Kentucky.

(CAD) system alerts safety officials that a high-priority incident is in progress and requires immediate response. The incident is logged into the CIMS and, depending on the extent and magnitude of the event, the local EOC may be activated to its appropriate operations level. All relevant agencies and departments are summoned to the EOC to prepare a coordinated response to secure public safety.

The EOC is equipped to accept maps and data feeds from all of the affected agencies into a common operating platform via an ArcGIS for Server installation. The EOC and the event partners can access the common operating platform via a GIS-powered situational awareness viewer. This environment lets the duty officer manage all aspects of the event in one common setting, or COP, ensuring that he or she has all the information necessary to make effective decisions.

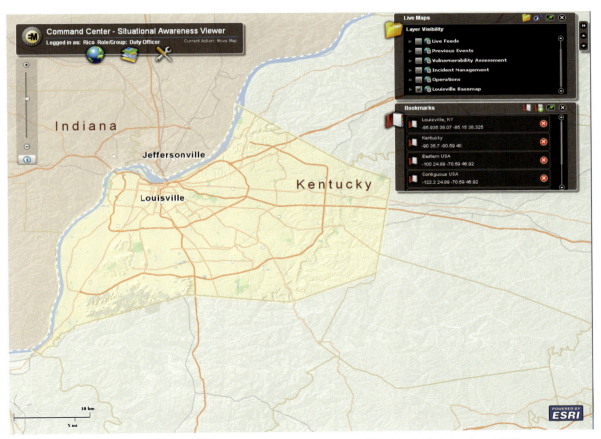

Command center situational awareness viewer COP Map by Esri; data from LOJIC, Louisville, Kentucky.

Once this event takes place, the managing officer quickly undertakes three tasks that effectively support the response and management of this critical incident:

- ◆ Determination of who and what is affected
- ◆ Determination of what actions to take to reduce the negative effects of the incident
- ◆ Communication of these actions to the appropriate officials

Using the ERG analysis tool in the situational awareness viewer, the managing officer or support personnel quickly model the affected area by entering the emitted material type reported by the police officer, along with the wind direction at the closest weather station, which is automatically populated by the live weather feed embedded in the system.

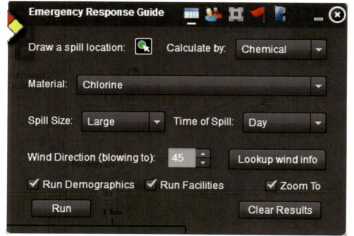

ERG chemical data entry screen
Esri.

The map viewer quickly characterizes the effect of the event by showing the areas of initial isolation and protective action, along with population and critical infrastructure (see page 19).

This is *actionable information* created by the GIS on the fly. It lets the officer determine approximately how many shelters to set up and what transportation, public utilities, schools, and other critical infrastructure might need to be shut down. This footprint is then used by the emergency notification system using the 9-1-1 database to notify everyone in both zones about the event and what steps they are to take to secure their safety.

Once the people in the area are informed and instructed regarding their safety, the second step is to reduce the negative effects of the

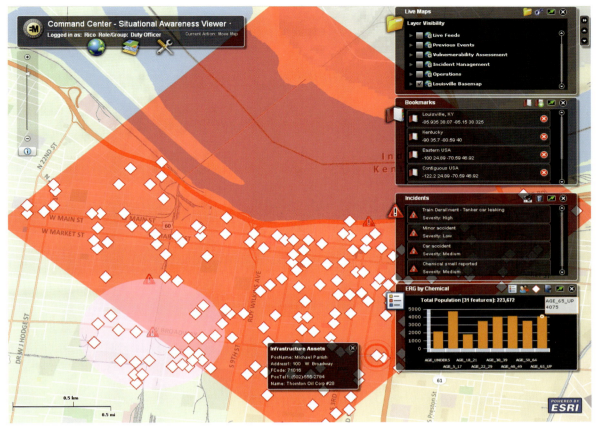

ERG map displaying initial isolation and protective zones and the affected population and critical infrastructure Map by Esri; data from LOJIC, Louisville, Kentucky.

spill. For this, the duty officer gets more help from the GIS through an additional set of advanced analytical tools available in the situational awareness viewer. A more refined plume of the potential spill is created using a more complex plume-modeling application such as ALOHA (Areal Locations of Hazardous Atmospheres), a system developed by the US Environmental Protection Agency (EPA) to simulate plume models.

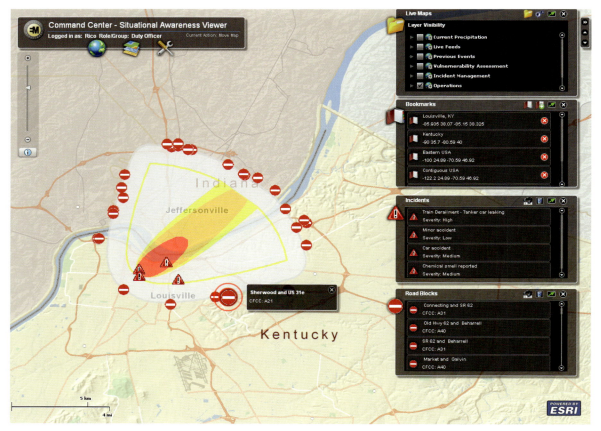

Command center situational awareness viewer displaying advanced plume model with delineated hot zones and roadblock locations   Map by Esri; data from LOJIC, Louisville, Kentucky.

The plume is buffered by a mile to delineate the hot zone. Roadblocks are set up around the perimeter to keep people out of the area. Shelters are set up outside the hot zone to relocate people who live in the affected area (see above).

Now that the first and second steps of the response phase have been completed, the managing officer communicates these actions to the appropriate officials and partner agencies. They are each given access to a secure, mission-specific viewer that contains briefing templates and information relevant to each ESF or presented with a digital or paper map for consideration. Each partner agency reviews its own information during briefing sessions to recommend further actions within its domain.

This scenario illustrates how effective the situational awareness viewer is in supporting a highly coordinated effort by participating agencies and response teams to quickly and effectively secure public safety in the most efficient manner. Chapter 5 provides a range of examples of how the geographic approach to emergency management provides effective planning and operations tools to secure public safety. The National Level Exercise is explored to illustrate the effectiveness of these tools.

These four workflow patterns make up the legs that support the ArcGIS common operating platform. Data management, planning and analysis, field operations, and situational awareness are what GIS does to systematically support emergency management workflow. If any one of these legs is broken, we lose the capability to provide a comprehensive systematic approach to emergency operations.

Chapter 6 is a guide to configuring key components of the ArcGIS common operating platform that deliver actionable information to managers and commanders. Selected templates and tools are explored that publish data embedded in the platform, bringing it to those making the critical decisions that save lives and protect property. Focus is on the implementation and configuration of key situational awareness viewers and map products using sample data provided from Naperville, Illinois. Configuration notes addressing how the viewers can be configured specifically for your jurisdiction are also presented to support the implementation of a system that enables the greatest situational awareness of emergency events in your own community. Best practices for ArcGIS for Desktop and ArcGIS for Server software configurations, hardware assemblies, applications, data, and resources to build the optimal system for effective emergency management planning and operations also are reviewed. This information is essential to the emergency manager and support staff implementing comprehensive situational awareness systems.

Chapter 2: Data management

"In the context of emergency operations, data management is gathering, managing, processing, and distributing information to users and across systems when and where needed. It is the capability to store, manage, update, and provide access to all of the unit's data through well-designed computer system architecture to meet the emergency management mission."[1]

One of the primary obstacles to successful emergency response and planning is the difficulty in gathering, managing, processing, and distributing the information needed to protect people and places. Robust data management lets an emergency management organization accurately plan, identify vulnerabilities, and provide comprehensive situational awareness by transforming the many disparate datasets coming from different agencies and portals into actionable intelligence. The GIS platform allows all types of data to be organized through its geographic location, making it available for analysis and operational use.

The GIS platform provides emergency management organizations with capabilities to support all aspects of their missions. Situational awareness can be achieved through a COP or multiple COPs that support specific missions. These might include viewers or COPs for command, operations, logistics, public information, or whatever mission is required. The platform thus becomes the common element, or the common operating platform, that enables mission-specific situational awareness.

1 Esri, "Geographic Information Systems: Providing the Platform for Comprehensive Emergency Management" (white paper [J9748], Esri, October 2008), 4.

GIS for data management   Esri.

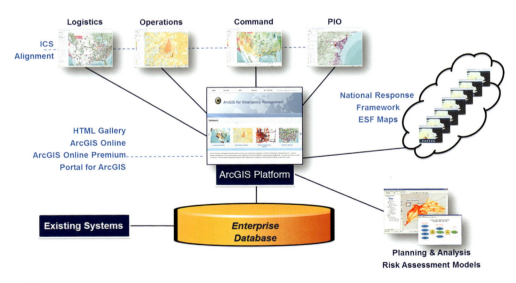

ArcGIS for emergency management mission–based situational awareness   Esri.

Collecting, storing, and managing data

Collaborating with partner agencies and organizations in the community to collect and access data can be one of the greatest challenges of an emergency management agency. The ideal emergency management data model integrates real-time and near-real-time databases from different agencies into a centralized common operating platform that can be assessed before, during, and after an event. In reality, however, such interoperability is not always easy. Agencies often have privacy issues and incompatible data formats, making data-sharing agreements sometimes difficult to broker. Agencies in different jurisdictions may also have difficulty sharing data because of incompatible data formats and inadequate file-sharing systems.

The GIS platform helps agencies overcome these interoperability problems by providing a stable hardware and software platform able to accommodate the great extent of detailed databases and high-resolution aerial imagery needed to be fully informed. The heart of this platform is the *geodatabase*, a central repository for data, specifically *spatial* data.

The geodatabase can store a rich and varied collection of relational spatial data and can be accessed by a single user through desktop software or by multiple users through an enterprise system across or between agencies.

Other strengths of the geodatabase as a data repository include its capacity to store the geographic location of each feature and the attribute information that describes it and its ability to apply sophisticated rules and relationships to the data. In the case of the overturned tanker, a first responder must have accurate location information for hydrants and water mains. A geodatabase of the relevant data would link the location of each fire hydrant to a parent water main in the system. The geodatabase would not allow a fire hydrant that is not spatially related to an adjacent water main to be stored in the database. This relationship rule ensures that the fire hydrants are located on the map in the correct position and that first responders can quickly find them on site when needed.

Geodatabase Esri.

A geodatabase can also store documents and plans that are linked to locations in the field, such as floor plans and photographs of buildings. These documents can also be managed within the GIS by their geographic relationship to features in the field, saving emergency managers valuable time during an event.

The geodatabase also has the capability to link agencies and operations active in an event, such as ICPs and support teams, regional and national agencies, and field personnel, including search-and-rescue operations. This integrated environment, visible as a display through a COP web-based map viewer with access to all types of documents associated with their location, enhances the situational awareness of all event partners connected through the geodatabase.

The foundation for all of this disparate information is the GIS common operating platform. It is the place where all data is collected, stored, and managed. This integrated platform houses the central geodatabase that enables data to flow into and out of the core to and from incident partners. Information and data models, available for download from the ArcGIS Resources website, provide a framework from which to construct a comprehensive geodatabase of a community's geography. Such models are designed by academic and industry leaders in association with Esri and are available across many industries. They present a best-practices schema for structuring geographic data and associated attributes into a functional geodatabase that forms the foundation of a well-designed GIS.

Local Government Information Model

The GIS data model recommended for emergency management and operations is the Local Government Information Model (LGIM). Because this model applies to the lowest level of civic operations, it contains the greatest number of detailed data layers used by governments and organizations to manage day-to-day operations. Thus, it is readily adaptable to higher levels of government and organizations, such as county, regional, state, and federal agencies, as they build comprehensive databases of geographic features and emergency resources under their jurisdiction.

The LGIM includes reference base data layers that define the geographic features of a community, such as parcels, streets, addresses, facilities, demographics, and infrastructure. In addition, this data model is enhanced by a set of emergency planning and operations data layers that address planned events and threats to public safety as they occur. These additional data layers include those relevant to emergency operations and public safety planning, such as access points, evacuation areas, incident locations, and emergency facilities. Coupled with local high-resolution aerial imagery embedded in the geodatabase, or accessed remotely from the ArcGIS Online basemap gallery, these layers in the LGIM form the local basemap and operational layers that characterize a community on a day-to-day basis, as well as the incidents and events that may pose a threat to public safety.

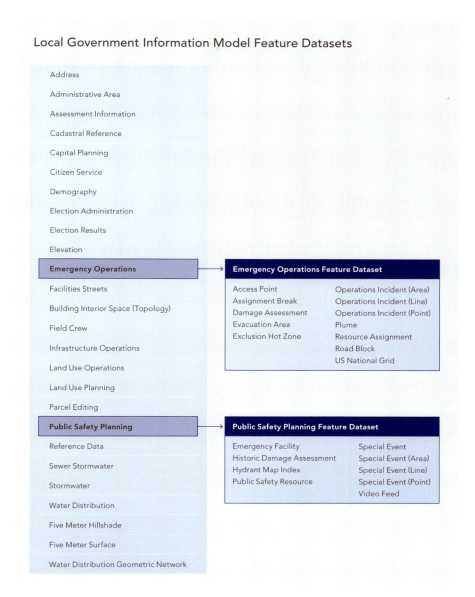

Local Government Information Model schematic Esri.

The schema, or empty shell, of the LGIM is available for download from the ArcGIS Resources Local Government—Public Safety Community website (http://resources.arcgis.com/en/communities/local-government/).[2] Local governments, including emergency operations managers, can use this schema to populate the model with their own local GIS data. By using this model, emergency managers are sure to include all relevant data layers in the GIS in a compatible format that aligns with best practices across the industry. This compatibility ensures interoperability with GIS data from other agencies, easing the difficulty often encountered when multiple agencies try to share data across jurisdictions and operating platforms.

LGIM basemap templates

Once the data is embedded in the geodatabase, it can be effectively used to inform decision making to save lives and secure property. This can only occur when maps depict accurately and communicate clearly where geographic features and events exist in the community. Well-designed operations maps begin with well-designed basemaps that tell the true geographic story of a community and event. Common symbology and cartographic conventions for features such as parcels, streets, and medical facilities help to ease the confusion and misinterpretation of where and what is occurring before, during, and after an event unfolds. Maps that can be read and understood by all personnel in the EOC and on site, within and between agencies, and across boundaries and jurisdictions go a long way to protect lives and secure public safety.

The Local Government Basemaps (LGBM)[3] is a set of templates that can be used to build basemaps from the LGIM that are essential to local government planning and operations. The basemaps provide important reference information that supports daily decision making. They orient map users and are typically combined with other map layers that represent operational information managed by a department or agency within local government.[4]

The LGBM includes the following basemaps:
- General Purpose
- Imagery Hybrid Reference Overlay
- Topographic

2 The Local Government Information Model, populated with sample data from Naperville, Illinois, is included in the downloadable templates configured in chapter 6.
3 Basemap templates are available for download on the ArcGIS Resources Local Government—Public Safety Community website: http://resources.arcgis.com/en/communities/local-government/.
4 "Local Government Basemaps (ArcGIS 10)," ArcGIS Resources Local Government—Public Safety Community, http://resources.arcgis.com/en/communities/local-government/01n40000003v000000.htm#GUID-04D78AAB-8CB6-455E-B41B-27679F998143.

- Parcel Public Access
- Public Safety
- Mobile Day
- Mobile Night
- Zoning
- Current Land Use
- Future Land Use

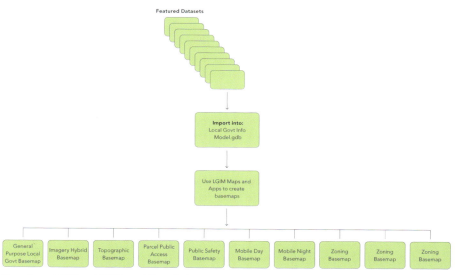

Data from LGIM to LGBM workflow Esri.

Of particular interest to emergency managers is the Public Safety basemap. This basemap provides a context map for operational data, such as incidents, events, and resources, that are added to the map as events unfold. It includes structures, roads, major facilities and land-marks, water features, parcels, addresses, and boundaries. The base-map uses design elements found in the General Purpose basemap, but emphasizes critical infrastructure and emergency facilities found in a community, such as major transportation arteries, hospitals, schools, police, fire, and emergency operations centers, and service locations.[5]

Collecting and storing the data that defines the geographic features of a community in a local government geodatabase ensures an integrated

5 "Local Government Basemaps (ArcGIS 10)," ArcGIS Resources Local Government— Public Safety Community, http://resources.arcgis.com/en/communities/local-government/ 01n40000003v000000.htm#GUID-04D78AAB-8CB6-455E-B41B-27679F998143.

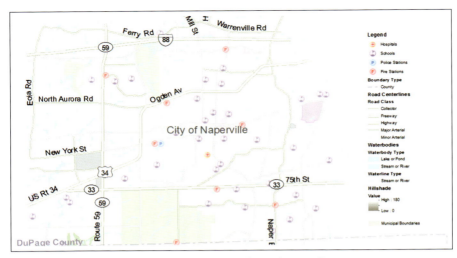

Public Safety basemap Map by Esri; data courtesy of City of Naperville.

and comprehensive starting point from which to manage operational data before, during, and after a significant event. The Public Safety basemap template built upon the LGIM data structure enables city managers and emergency operations personnel to have the same information drawn from the same data source using the same cartographic elements and styles, ensuring compatibility and consistency between agencies and organizations.

LGIM operational layers
The LGIM includes two additional feature datasets: one that pertains to public safety planning and one to emergency operations.

Public safety planning
The Public Safety Planning dataset is a collection of features gathered during public safety planning and preparedness activities. It is essentially the basemap of public safety planning, pinning down the location of existing emergency services, facilities, and resources before an incident occurs. The Public Safety Planning dataset includes the following feature classes:

- *Emergency Facility*—The location of emergency facilities and status and point of contact information. This includes the location of the EOC for the region or organization.
- *Historic Damage Assessment*—The location of damaged structures and a description of the extent to which each structure was damaged. This provides a baseline from which to measure the extent of damage from a current event.
- *Hydrant Map Index*—The index grid used to create the fire hydrant map book used in fire response activities.

- *Public Safety Resource*—Logistics resources or skills—such as personnel, teams, facilities, equipment, and supplies used during an emergency incident or event—inventoried in a system, such as the National Incident Management System (NIMS) Incident Resource Inventory System (IRIS).
- *Special Event*—Description of special events in the community that require the allocation of public safety resources and assets.
- *Special Event (Area)*—The location of polygonal assets (parking, VIP area, restricted access, etc.) allocated during a special event.
- *Special Event (Line)*—The location of linear assets (vehicular and pedestrian routes, etc.) allocated during a special event.
- *Special Event (Point)*—The location of individual assets allocated during a special event.
- *Video Feed*—Closed-circuit video feeds coming from traffic, surveillance, and other sources.

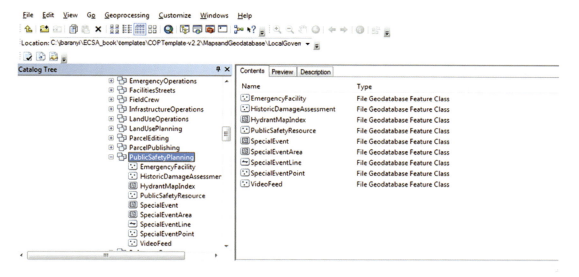

Public Safety Planning feature dataset Esri.

Preparedness depends heavily on having an accurate record of existing public safety resources. Access to emergency resource inventories serves to expedite response to a potentially catastrophic event. It is important to know what logistics resources are available to emergency personnel if and when such an event occurs. These resources, however, may not be useable if an emergency manager doesn't know where they are located and how close they are to an incident and the people who need them.

Locating where response resources are housed identifies areas that may not have such resources available to them if and when an event occurs. In this case, emergency operations planners can use the integrated

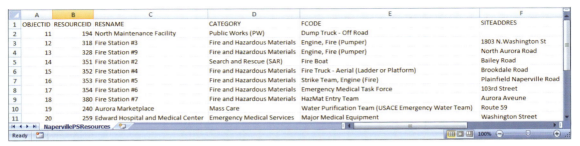

Public safety resources inventory table Esri.

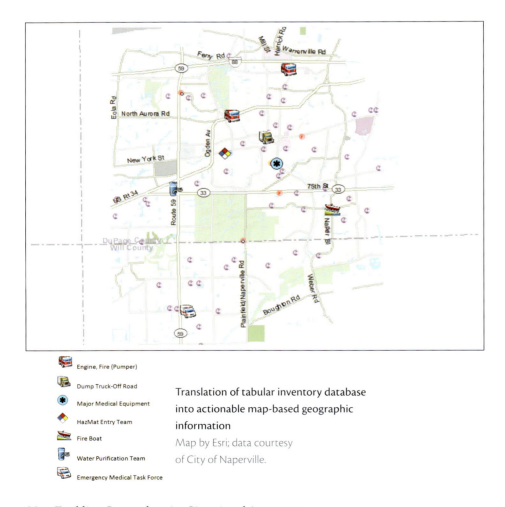

Engine, Fire (Pumper)

Dump Truck-Off Road

Major Medical Equipment

HazMat Entry Team

Fire Boat

Water Purification Team

Emergency Medical Task Force

Translation of tabular inventory database into actionable map-based geographic information

Map by Esri; data courtesy of City of Naperville.

geodatabase to inform plans and broker agreements in the pre-event planning phase, calling upon neighboring jurisdictions for support, if necessary. For example, in the case of the overturned tanker, converting lists of available logistics resources, such as generators, potable water, mobile feeding kitchens, and debris removal teams and trucks, into features on the map that can be queried and analyzed geographically to support a response is more effective than trying to determine from traditional spreadsheets if these resources are located in close proximity to where the event occurred.

The managing officer on hand can see on the map where resources are located throughout the region, within the area of concern as well as outside the current jurisdiction, and can request resources using sharing agreements that have already been brokered between agencies during the planning phase to accommodate immediate supply demands.

Public safety resource inventory

The Kansas Adjutant General's Department, Division of Emergency Management (KDEM), has developed an efficient system for managing and tracking public safety resources across the state. Initially, each county was provided a template in spreadsheet format to list and type each of its public and private public safety resources according to the NIMS Tier I and II resource-typing protocol. These include such resources as law enforcement, medical, and search-and-rescue teams; heavy equipment, such as backhoes and concrete cutters; and air and ground ambulances. The spreadsheets were returned and integrated into the department's geodatabase, where each resource location was then geocoded with a set of coordinates to be placed on the map. This setup has migrated to a new web-based resource manager system deployed to enable participating counties and other jurisdictions to enter and update their resource inventories online. The data from this system is continuously served over the Internet and integrated into the department's geodatabase, ensuring that it is always up to date (see figure on page 34).

Resources stored in the KDEM geodatabase are queried during the planning of a special event or during an emergency operation and subsequently assigned and deployed, as needed, to meet the safety needs of the community. Through the use of its Find Closest Resource tool embedded in the online COP application, the Kansas Mapping Application Program (Kansas-MAP), the department's resource inventory becomes actionable information. By entering the address of a planned event or emergency incident, the system generates a list of the closest resources queried; shows the route on the map, along with driving directions, detailing the quickest way to get to the resource; and provides contact information on file for the desired resource. Through the integration of this type of actionable information into the public safety planning process, emergency managers have the greatest situational awareness of available resources in the event of a catastrophic incident.

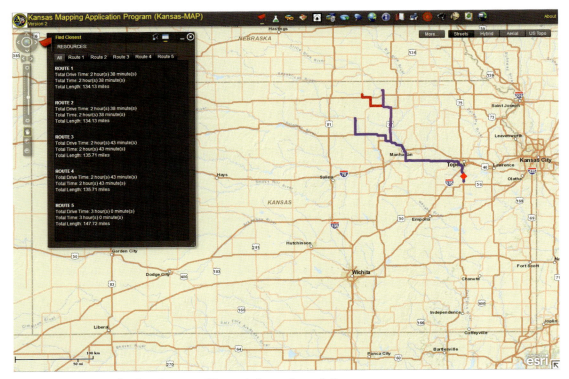

Kansas-MAP COP application Courtesy of Kansas Adjutant General's Department, Division of Emergency Management.

Emergency operations

The Emergency Operations dataset contains a collection of features used to gather information about dynamic emergency incidents or planned special events:

- *Access Point*—The location of emergency access points.
- *Assignment Break*—The emergency incident resource assignment breaks.
- *Damage Assessment*—The location of damaged structures and a description of the extent to which each structure was damaged. This data may come from a field operations damage assessment crew supplied with mobile GIS and mobilized during the recovery stage of an event to record an inventory of the postevent condition

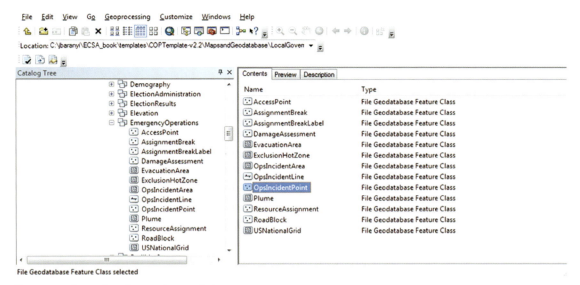

Emergency Operations feature dataset Esri.

of each parcel and structure in the community. This information is critical when applying to FEMA for federal recovery funds.

◆ *Evacuation Area*—The evacuation area caused by the emergency incident. This information may result from a spatial analysis using readily available analytical tools that quickly evaluate the geographic extent of the affected area or from more advanced tools using scientific analysis to determine who may be affected and how.

◆ *Exclusion Hot Zone*—The area immediately around a spill or release where contamination can occur.

◆ *Operations Incident (Area/Line/Point)*—The affected area, line, or point for an incident.

◆ *Plume*—Plume polygon features generated from advanced modeling tools, such as the EPA ALOHA plume-modeling tool that can be used with ArcGIS software.

◆ *Resource Assignment*—Resources assigned during an emergency incident or event. This information may derive from the Public Safety Planning feature dataset described above, where resources and skills are inventoried in a system such as NIMS IRIS and then queried using a tool, such as the Find Closest Resource tool, to assign resources to meet the safety needs of the community.

◆ *Road Block*—Current roadblocks with affected cross streets.

◆ *US National Grid*—Alphanumeric reference system that overlays the UTM coordinate system.

This dataset of the LGIM geodatabase remains inactive or in passive observation mode until an event occurs that requires an emergency response. Once an event is escalated to an emergency response level, the live feeds of incident data that flow into the geodatabase provide all partner agencies active in an event with a common base to work from where the operational data is current and accurate at all times. Also, because the preparedness data is also housed in the same geodatabase, it can be readily accessed and made operational as the event unfolds, enabling an effective allocation of emergency supplies and equipment to the right people and places at the right time.

Operations incident data feeds

The primary focus of this feature dataset is the operations incident data feeds, which include continuous information from business systems within the organization, such as a CIMS or CAD police, fire, and ambulance 9-1-1 systems. An emergency manager is most likely to be notified of an emergency event through these data feeds. Once the incident is logged into the CIMS and is visible on the map as an active event, the emergency manager activates the EOC and puts its resources in motion to quickly respond in a well-coordinated and effective way. As the event unfolds and first responders are on site, the data feed continues to update the map with the most recent conditions in and around the affected area.

An innovative system that integrates dynamic data from live incident feeds has been developed and deployed by E-Comm 9-1-1, the public safety answer point (PSAP) for southwest British Columbia, Canada. E-Comm 9-1-1 also dispatches for thirty police and fire departments across the region and manages and supports a consolidated wide-area radio system that allows police, fire, and ambulance personnel to communicate across jurisdictions.

The integrated incident management system is built upon a multifunctional web application, the Emergency Event Map Viewer (E2MV), which enables live incident data from multiple agencies to be viewed in a single map viewer, facilitating information sharing across multiple dispatch agencies.

This GIS/CAD data integration application is designed to provide a COP for emergency services and is also capable of connecting to other spatially enabled data systems used by utilities, transit, and public works agencies. The system architecture includes the front-end web-mapping application, which combines all of the data feeds into one map display. This application is built on an ArcGIS platform and fed by a

Vancouver E2MV Courtesy of Emergency Communications for Southwest British Columbia Incorporated.

Representational State Transfer (REST)–based web service[6] (Emergency Event Web Service [E2WS]), which includes a data importer that connects and collects information from the external systems of event partners. The live connection to police, fire, and ambulance events is refreshed every 60 seconds or sooner, depending on the event.

It was initially designed in 2007 to address a lack of interoperability between public safety partners in the region. The E2MV mapping system is integral to geospatial decision making between E-Comm 9-1-1 and its partners in public safety. It also supports a connection with the City of Vancouver Emergency Operations Centre, which is challenged with monitoring the day-to-day public safety operations of a large metropolitan area, as well as coordinating major public events that require coordinated preparation and response to secure the safety and security of the city.

The greatest challenge to the system was to adapt it to the safety needs of the 2010 Vancouver Winter Olympics. Multiple public safety partners beyond the normal scope of the agency's reach, including thirteen police

---

6 **Representational State Transfer (REST):** RESTful web services are web services that transmit data over HTTP (Hypertext Transfer Protocol) without an additional messaging layer, such as SOAP (Simple Object Access Protocol). In the most common RESTful web service architecture, the client sends all parameters in the request URL (uniform resource locator). Pinde Fu and Jiulin Sun, *Web GIS: Principles and Applications* (Redlands, CA: Esri Press, 2011), 63.

agencies, seven fire departments, and one ambulance service, were successfully integrated into the system through the development of comprehensive data-sharing agreements. The system architecture was able to import the live data feeds from these agencies without difficulty, and the Olympic security force was able to effectively use the system's COP to monitor live activities in and around the multiple event venues at all times throughout the region.

Vancouver EOC Photo courtesy of Emergency Communications for Southwest British Columbia Incorporated.

The post-Olympics success story continues as E-Comm 9-1-1 returns the system to support basic everyday operations, as well as major public events regularly scheduled in the Vancouver area. Many of the data-sharing agreements brokered in support of the Olympics have been extended to include the additional public safety agencies into the regular live incident data feed system, enhancing the scope and breadth of the E2MV system through the southwest region of British Columbia. Additional agreements are in the works to link this system with other provincial and federal initiatives, including the Integrated Border Enforcement Team and Joint Rescue Coordination Centre.

Other data supporting the LGIM geodatabase

The LGIM geodatabase is one piece of the common operating platform in service at an emergency management agency or EOC. Additional data and information that support and extend the LGIM geodatabase include other relevant databases, live data feeds, documents, and high-resolution imagery that make up the complete planning and operating environment of an emergency management organization.

High-resolution imagery

The use of high-resolution aerial and satellite imagery in emergency management planning and operations is critical to the success of a mission. High-resolution imagery offers emergency response teams opportunities to enhance disaster documentation and future planning, ground control, and guidance for emergency response teams; evacuation planning; hazardous spill detection; postdisaster damage survey; and disaster recovery planning.[7]

Imagery comes in a variety of formats. Satellite imagery is acquired on an ongoing basis from both government and civilian satellites, such as Landsat, SPOT, QuickBird, IKONOS, and OrbView/GeoEye 1. Pixel resolution (how much land or water each pixel on the image covers on the ground) ranges from low-resolution images at 1 km² to the highest resolution images captured at submeter levels. Low-resolution images are best to show regional patterns of land use and environmental change over wide swaths of the earth, such as weather systems and damage from major earthquake and flooding events. High-resolution imagery showing before and after street-level detail of an affected area is invaluable to ground-level response and recovery teams navigating through communities in the aftermath of an event (see figures on page 40).

Spectral resolution (the wavelength interval in the electromagnetic spectrum that a scanner can "see") ranges from simple images that show everything the human eye can see to scenes captured beyond the visible spectrum. Updated high-resolution images within the visible range essentially put emergency managers on the site while still physically at the EOC, enabling them to get a better understanding of where the damage is. In addition, thermal infrared scanners capture images in that portion of the spectrum where emergency managers and responders can't see beyond the visible range to view the extent of wildfires and plume events as they unfold on the ground and in the air.

Airborne imagery—imagery captured not from satellites, but from sensors and cameras mounted on aircraft typically flying at an altitude of between 2,000 and 12,000 feet—provide emergency managers with an additional source of detailed information. Many jurisdictions task aerial surveys to update land use and master planning activities at regular annual or biannual intervals. During major catastrophic events, Civil Air Patrol provides aerial survey services to capture imagery concentrated on an affected area or a location where search-and-rescue operations are deployed. These

---

7 "The Use of Air Photos in Emergency Management," February 1, 2001, http://www.husdal. com/2001/02/01/the-use-of-air-photos-in-emergency-management/.

GeoEye high-resolution satellite images of Sendai, Japan, before and after the 2011 tsunami

Aerial imagery captured from low-flying aircraft before and after the 2009 bushfires in Victoria, **Australia** Courtesy of Department of Sustainability and Environment, the State of Victoria, Australia.

high-resolution images are invaluable sources of critical information to an emergency response team that must have complete situational awareness of a place under siege to secure its safety (see figures on page 41).

Although this vast array of imagery provides a rich collection of vital information necessary for successful emergency response, collecting, storing, and managing such data can pose a colossal challenge to an emergency management organization. During the 2007 wildfires in California, the US Department of Homeland Security (DHS) provided numerous assets that collected large volumes of data—infrared imagery, high-definition video, satellite imagery, and much more; however, no effective data management system was in place, so actionable information was difficult to obtain and deliver to incident management personnel when they needed it most.

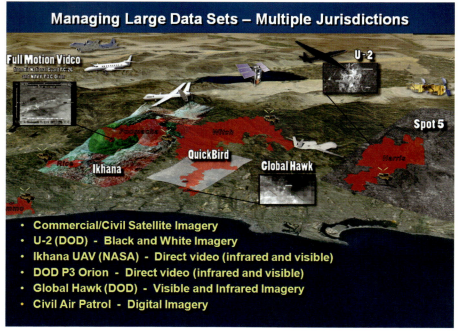

Managing large datasets during an emergency event Figure by Esri; Cal Fire Siege data courtesy of David Blankenship, 2007, FireSiege, SouthOPS; satellite imagery from Esri Data & Maps, courtesy of Earth Satellite Corporation.

The importance of a well-built common operating GIS platform capable of storing and managing such vast amounts of imagery and data becomes even more evident in light of the pressing need for high-resolution imagery. ArcGIS provides an integrated platform where basemap and operational data stored in the LGIM geodatabase are geographically aligned with the extensive collection of relevant imagery made available from various agencies and organizations before, during, and after an event. In addition, the ArcGIS platform integrates seamlessly with the rich collection of online high-resolution imagery services offered in the ArcGIS Online data library. All of these data and imagery resources can be effectively integrated within this platform to provide both a common view and a range of mission-based perspectives of the information over multiple environments (i.e., desktop, web, mobile), enabling the greatest degree of situational awareness available to all partners in an emergency response.

The Florida Division of Emergency Management's (DEM) Geospatial Assessment Tool for Operations and Response (GATOR) is built upon the ArcGIS platform and integrates a vast collection of basemap and operational data with local and online imagery products and services. This interactive web-mapping tool offers displays of geographic information that support emergency preparedness, operations, and response, providing better situational awareness for the DEM's State Emergency Operations Center (SEOC).[8] GATOR was used extensively during Florida's response to the Deepwater Horizon oil spill off the Florida coast in the Gulf of Mexico. Because the State of Florida has a very progressive open public records law, up-to-date high-resolution imagery from counties, water management districts, and state agencies is available from Florida State University and made available

Florida disaster GATOR map viewer Courtesy of Florida Division of Emergency Management.

8 "DEM News," Florida Division of Emergency Management, December 17, 2010, http://www.floridadisaster.org/documents/2010%20Winter%20Newsletter%2020101217.pdf.

to map services operated by Esri on the ArcGIS for Server platform. GATOR is able to leverage this imagery service through the ArcGIS Online basemap gallery, simplifying the storage and access of such a vast collection of data. The DEM does, however, store within its ArcGIS system nearly a terabyte of data, including the most up-to-date high-resolution imagery available for each county, as well as the most recent statewide National Agriculture Imagery Program (NAIP)[9] imagery as backup in case of emergency when connections break down and access to online web services is disrupted.

An unexpected benefit of aerial imagery during the Deepwater Horizon response involved the use of geotagged airborne photos acquired by the Civil Air Patrol. Over 85,000 low-altitude images were captured throughout the region to initially monitor shoreline vegetation restoration and oil boom placement. As the geotagged images were placed

The use of GATOR to monitor boom operations via geotagged images during the Deepwater Horizon oil spill recovery operation Courtesy of Florida Division of Emergency Management.

9 "The National Agriculture Imagery Program (NAIP) acquires aerial imagery during the agricultural growing seasons in the continental United States. A primary goal of the NAIP program is to make digital orthophotography available to governmental agencies and the public within a year of acquisition. NAIP is administered by the USDA's Farm Service Agency (FSA) through the Aerial Photography Field Office in Salt Lake City. This 'leaf-on' imagery is used as a base layer for GIS programs in FSA's County Service Centers, and is used to maintain the Common Land Unit (CLU) boundaries." US Department of Agriculture, Farm Service Agency, http://www.fsa.usda.gov/FSA/apfoapp?area=home&subject=prog&topic=nai.

on the map, DEM personnel began to notice that although the mission was to focus reconnaissance along the shoreline and out into the Gulf to monitor boom placement, geotagged images were showing up inland, away from the expected areas of interest. It became apparent that by capturing images of boom staging areas as supplies came into rail depots in the region emergency operations planners could estimate how many feet of boom were available in the area that could be deployed to hold back the movement of oil toward the coastline.

This use of current geotagged aerial imagery leveraged the versatility of georeferenced information embedded in an integrated, well-managed, and well-maintained common operating GIS platform. By expertly storing, cataloging, geotagging, and managing this imagery dataset, Florida's State Emergency Response Team (SERT) was able to effectively deliver it to emergency planners and operations personnel who were then able to make effective decisions about boom placement that would secure and protect the people and places affected by this significant environmental event.

Planning and operations documents

Planners and emergency operations personnel spend a lot of time and expertise compiling preparedness and mitigation plans that outline emergency procedures designed to ensure the safety of people and places in the event of a catastrophic event. These plans include documents outlining standard operating procedures (SOP), Geospatial Concept of Operations (GeoCONOPS),[10] building floor plans, security video clips, evacuation procedures and routes, and photographs of buildings and details about their pre-event conditions.

Preparedness documents and plans are difficult to access and put to use if paper copies remain in binders and file cabinets in civic offices and EOCs. By dynamically linking digital copies of these documents to features in the LGIM geodatabase, planners and operations personnel can quickly and effectively access these documents at all times with the simple click of the mouse on the map display. These critical documents can be put to use as an event unfolds to provide immediate actionable intelligence, enabling emergency managers to coordinate a well-planned response that saves lives and secures property.

In the case of the overturned tanker, the affect that this event can have on elderly residents living in nearby nursing homes can be considerable. To secure their continued safety, they must be either

10 **Geospatial Concept of Operations (GeoCONOPS)**: The Federal Interagency GeoCONOPS is intended to identify and align the geospatial resources that are required to support the National Response Framework, Emergency Support Functions, and supporting federal mission partners. "Federal Interagency Geospatial Concept of Operations (GeoCONOPS)," June 2008, DHS_ GeoCONOPS_v1.0_8.5x11.pdf.

sheltered in place or evacuated to designated shelter locations outside of the affected area. Having ready access to the response plan for the affected facility directly from the map display by clicking the nursing home feature on the map significantly decreases the amount of time and coordination necessary to deploy the evacuation.

Responders can swiftly gain access to the site by reviewing the attached site plan and referring to the floor plan and photographs of the interior spaces to determine the most efficient way to secure the safety of the residents. These documents are an important piece of the common operating GIS platform, quickly providing emergency managers and first responders with actionable information directly aligned with the geographic location of the facilities and residents at risk.

Other live data feeds

Additional important data that supports the LGIM geodatabase includes real-time live feeds from weather sensors, traffic webcams, and systems that detect and monitor such things as wildfires, earthquakes, tsunamis, hazardous materials, and available shelter locations across a jurisdiction. This data represents a vital link to continuous dynamic information fed into the GIS to sustain a complete situational awareness of an event. The data may be connected to the operating platform through a web service data importer, as described in the Vancouver system discussed earlier, or it may be integrated via an RSS widget that embeds a live data feed from an outside agency directly into the map display.

RSS feeds, or Really Simple Syndication, are frequently updated text or image information that is automatically accessible to users who subscribe to the feed directly through their browser or from a website that incorporates the feed into its content. Many web users consume this type of data to remain up-to-date on current news, sports scores, travel status, and weather conditions. By clicking on an RSS link[11] in a browser or web page, a user is supplied with a string of text containing the latest information about an event, such as the score of a current NHL hockey game, or directed to a web page containing the full news story about a particular event.

A GeoRSS is an enhanced data feed that not only provides up-to-date information about a particular feature or event of interest, but also includes a geographic tag that locates it on a map. It attaches geographic coordinates to photos, text, and other digital information that is available on the Internet. GeoRSS feeds that support emergency management planning and operations include those that plot the locations of conditions significant to the safety and well-being of a community. The US

11 RSS logo courtesy of the Mozilla Foundation.

Geological Survey (USGS) earthquake updates and the National Oceanic and Atmospheric Administration (NOAA) Incident News website provide GeoRSS feeds that can be directly embedded into a map viewer.

Other live feeds, such as weather feeds from NOAA and the National Weather Service (NWS), include current conditions and forecasted warnings of severe hurricanes and cyclone events. The locations of current worldwide seismic events and nationwide river conditions are continuously updated by the USGS's Natural Hazard Support System (NHSS) feed, which is based on the ArcGIS platform, and embedded in the template used to program a COP situational awareness viewer. Feeds from the Pacific Disaster Center (PDC) also are now served from the ArcGIS platform and embedded in the template.[12]

Earthquake GeoRSS feed embedded in a situational awareness map viewer
Esri.

The Ventura County Fire Department (VCFD) integrates an extensive set of live data feeds into its web-based decision support system. The Situation Analyst (available from Intterra) system uses the ArcGIS common operating platform to draw dozens of live data feeds from multiple sources into its map viewer. Information related to wildfires, hazardous materials, flooding, other natural and human-made disasters, and more are supported by the integrated ArcGIS platform.

---

12 Natural Hazard Support System, http://nhss.cr.usgs.gov; Advanced Hydrologic Prediction Service, http://www.nws.noaa.gov/oh/ahps; Pacific Disaster Center, http://www.pdc.org/.

Ventura County Fire Department's use of live data feeds in its web-based decision support system
Courtesy of VCFD Situation Analyst.

The VCFD system is designed to import GeoRSS, file geodatabases, shape-files, and other file types into its system on the fly as information becomes available from other sources during operations, and then reformat them as REST services to be consumed by other users and applications. This capability not only enhances the situational awareness the VCFD has of its domain, but it enables interoperability among hundreds of other agencies, users, media, and the public who depend on the system for real-time operational updates before, during, and after an event unfolds. The VCFD Situation Analyst deployment is designed as an open and collaborative system; it is interoperable with the Los Angeles Fire Department LASER (Los Angeles Situation Awareness for Emergency Response), CAL EMA (California Emergency Management Agency), and USFS (US Forest Service) Enterprise Geospatial Portal decision support systems. It leverages data and services from each and is built upon a cloud architecture, thus ensuring availability during regional disasters of any kind. For the VCFD Wildland Fire Support Mission, Situation Analyst integrates live, continuous overhead thermal infrared support from the Range and Bearing AWIS sensor. These functions allow VCFD operations personnel to edit the mapping in the field, receive live infrared from the aircraft (as a REST service), and publish an official perimeter with the click of the mouse, enabling the VCFD to continu-ally manage dynamic organizational knowledge and drive the public message

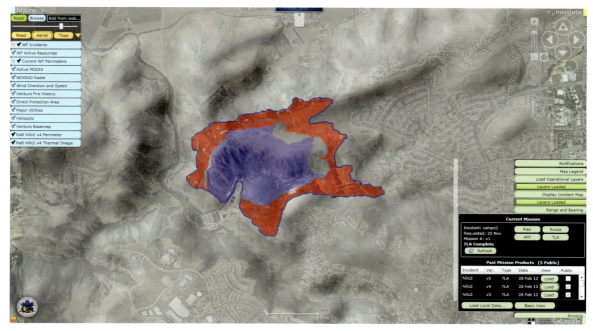

Ventura County Fire Department's use of live thermal infrared imagery as a web service

Courtesy of VCFD Situation Analyst.

instead of reacting to media depictions of the event. The system is modular and continues to expand and fold in more and more department workflows.

Social media and volunteered geographic information

Volunteered geographic information (VGI), generated through the use of social media, is another source of relevant and dynamic data important to sustaining comprehensive situational awareness of a community or an event. Similar to live data feeds, discussed above, VGI is georeferenced information that derives from a place and provides information about the conditions at that location. The difference is that this information is not sourced from an official agency or organization, but rather from local citizens sending information out into the worldwide cyber network via social media. SMS (Short Message Service) and platforms such as Facebook, YouTube, Flickr, and Twitter provide opportunities for anyone with an Internet-ready device and user account to share personal text and images linked to location with the global digital community.

These data feeds are streamed within the web environment of the application where "friends" and "followers" can read and share the content. They may contain a personal comment about an incident, a photograph or video of an event as it unfolds, or a link to a published article or press release from a mass media outlet or local publication. More compelling, however, is when

Ushahidi map of crowd-sourced reports from Haiti after the 2010 earthquake

this disparate content is compiled and curated through a single portal, such as Ushahidi, or one of Esri's online disaster response maps. This type of display merges all activity from all devices and channels into one crowd-sourced environment. When placed on a map, this convergence of VGI shows the pattern and magnitude of activity around a common place and time.

Esri and Ushahidi have also collaborated on an add-in tool that enables users to bring Ushahidi data into ArcGIS where analyses can be performed to extract even greater actionable information from crowd-sourced data points. For example, points representing the location of people reporting current site conditions along inundated waterways in Thailand during a 2011 flood event can be further processed to show statistically significant clusters of reports in areas requiring immediate assistance.

Within the context of emergency management, this data stream essentially multiplies the sensors on the ground that gather site-based information beyond the capabilities of existing scanners and cameras. In the case of the overturned tanker discussed earlier, a fire at the site may disrupt power and landline data communications, or the site may be in an area where communications are limited. A field commander may not be able to depend on traditional sources of initial information to make early decisions about where best to deploy first responders. Accurate crowd-sourced information from people in the area may

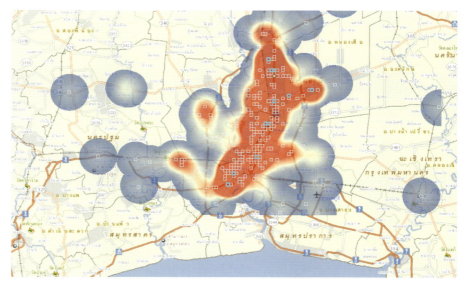

Ushahidi2ArcGIS tool used to process crowd-sourced point data into concentrations of VGI reports © 2012 Ushahidi, Inc.; OpenStreetMap and data licensed via CC by SA 2.0.

provide an emergency response team with immediate on-site information about the ground conditions at a site before an official assessment has been made. In essence, these people become the "first" first responders to an event. For example, during the emergency response to the Japan earthquake and tsunami, crowd-sourced data from social media about power outages were mined, and then clusters of data feeds were translated into hot spots on a map to indicate the areas where safety concerns were greatest. This type of data management can reduce the standard operational map update time from 12 hours to hourly, making a huge difference in the effectiveness of a response, and ultimately going a long way to protect lives and secure property as quickly as possible.

The effect that this type of communication has had on world events over the past three years has been unprecedented. Political upheavals in Iran, Egypt, Libya, Kenya, and the Sudan, as well as emergency recovery efforts along the Gulf Coast in the United States and in Haiti and Japan, depended heavily on real-time text messaging, Tweets, Flickr photos, YouTube videos, and Facebook status updates, many verified, compiled, and mapped on the Ushahidi and Esri platforms.

Integrating VGI into a GIS does, however, pose some considerable risks. As noted, "in contrast to GIS-based data, which is organized with consistent data models and collected systematically, VGI is mostly observational and qualitative, and very rarely is it collected systematically in a science-based framework.

Typically, it is not collected in a structured framework, nor is it associated with metadata, and there is little or no formal responsibility for data quality."[13]

Nevertheless, it is possible to incorporate this data into a GIS such that it can add to the integrity of the geodatabase beyond visualization. Simply linking observational data, such as texts and photos offered by local citizens, to existing features in a geodatabase, such as building footprints or water bodies, enhances the attribute, or descriptive, information about those features without compromising the geographic integrity of the data. Mining social media data by "at signs" (@) and "hashtags" (#) enables quick and effective extraction of only those postings relevant to a specific event. For example, putting an @ before the Twitter ID of local emergency response agencies, news outlets, or other relevant people or organizations (e.g., @femaregion7) or adding a hashtag that encodes a searchable keyword in the message streams requests and responses directly to the relevant agency (e.g., #Joplin) and filters out any postings not related to the tornados in Joplin, Missouri, located in FEMA Region VII.[14]

Providing web-based data entry interfaces where citizens can report conditions after an event occurs can supplement the damage assessment feature of a geodatabase. In such an environment, accuracy can be improved by providing predefined digital forms where users answer structured questions that align with the feature class of the database. For example, a citizen may report the condition of nearby homes after a gas explosion by publishing geotagged Tweets, photos, and videos of the neighborhood. These items are then linked to the building footprints of the homes, thereby accurately incorporating this important social media data into the full set of data that supports the geodatabase. In addition, that same citizen can add structured information via a digital form on a web or mobile app that asks specifically for the condition of the roof, windows, and surrounding property. This VGI data is entered into the geodatabase in a controlled format that complies with the standard framework as defined in the LGIM geodatabase. It can provide a first-glance view of damage assessment before an official crew can get on site.

Hamilton County, Tennessee, developed an application whereby residents can report storm damage to their property via an online web tool. This system was quickly deployed in response to the severe storms and tornado outbreak that struck the southeast United States in the spring of 2011. The map-based application asks for the street address of the property, and then allows the

---

13 Matt Artz, "Structure: The Key to Volunteered Geographic Information—New Tools Enable the Public to Participate in GIS Database Development," *ArcWatch,* March 2010, http://www.esri.com/news/arcwatch/0310/volunteer.html.

14 See the "Twitter Hashtags and Emergency Management" post on the *Urban Areas Security Initiatives* blog for a list of popular hashtags used in emergency management and operations: http://urbanareas.org/blog/?p=150.

user to select from a drop-down list what type of damage has occurred, including structural damage from a tree, or wind or tornado damage, affecting 10 to 100 percent of the property. Each entry is plotted on the map with an indication of whether the entry was self-reported or entered by the county during an official damage assessment survey. The Hamilton County Government is soliciting this information to help assemble the most complete and accurate data for mapping of all storm-related structure damage.

Hamilton County, Tennessee, public information web portal for volunteer-generated damage assessment information Courtesy of Hamilton County GIS.

Entering this information into a predesigned form ensures that the data is consistent with the integrity of the geodatabase used by the county to plan and manage its response to a major weather event. This type of VGI, in conjunction with free-flowing posts from the various social media platforms, provides a rich source of valuable and pertinent information that enhances the overall awareness of communities and events that occur within them.

The ways in which everyday citizens interface with the world will undoubtedly evolve and change over time. One thing is certain, however, and that is that people will make a greater and greater contribution to geographic information as technology becomes more available, data more democratized, and pathways between the two more operational. Enriching our geodatabases and map displays with VGI that is processed within a well-defined and maintained GIS data management system enhances our situational awareness of communities and events, ensuring the safety of people and places should a disaster or catastrophic incident occur.

Data sharing

As noted throughout this discussion, one of the greatest challenges to designing a sound and effective GIS data management strategy is collaboration among event partners and sharing of relevant data before, during, and after an incident occurs. Catastrophic events, particularly natural disasters, know no boundaries and are not generally obstructed by political borders that separate jurisdictions. To meet these challenges, virtual "communities" are emerging as online meeting places, hosted in the cloud, where people and partners who share common interests and events can collaborate.

ArcGIS Online

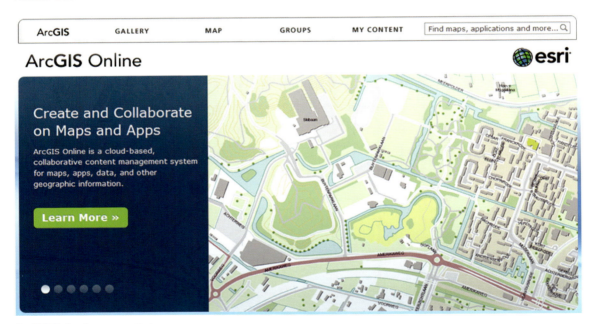

ArcGIS Online homepage Esri.

ArcGIS Online is a complete, cloud-based, collaborative content management system for working with geographic information. The platform provides an on-demand, secure, open, and configurable infrastructure for creating web maps; web-enabling data; sharing maps, data, and applications; and managing content and multiple users from within or across agencies and organizations. It includes 100,000 maps, data, configurable templates, GIS tools, and APIs (application program interfaces) for developers and custom applications published by the GIS community, including Esri, the USGS, the US Census, the World Bank, and local governments and agencies around the world. A subscription to ArcGIS Online gives users and organizations enhanced capabilities for publishing data to Esri's cloud and configuring a website to manage content and users.[15]

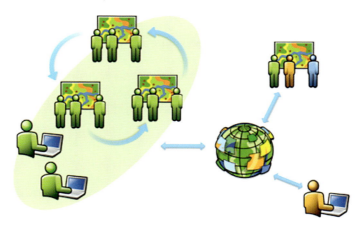

ArcGIS Online workflow Esri.

The advantage that the ArcGIS Online environment affords public safety and emergency management practices affects the administration and delivery of critical time-sensitive information. The extensive directory of active map services provides agencies and organizations with data that supplements their own base layers contained in the LGIM, reducing the extent of downloading, processing, and managing of in-house datasets needed to provide full situational awareness of a community. Accessing readily available geospatial information found on ArcGIS Online, such as topographic imagery, shaded relief, ocean, reference, and transportation layers, can be immediately incorporated as a foundation for mapping operational layers.

The delivery of real-time relevant information across this platform at any given time and place, among the ArcGIS Online community at large or within selected user groups targeting particular industries or events, provides the ideal integrated architecture to host a coordinated response and recovery operation. Multijurisdictional partners managing the public safety of a community are particularly well positioned to host shared work

15 Adapted from "ArcGIS Resources: What Is ArcGIS Online?" February 2, 2012, http://help.arcgis. com/en/arcgisonline/help/index.html#/What_is_ArcGIS_Online/010q00000074000000/.

groups in this online cloud community, where each participant enables other users in the group to view maps and integrate analysis tools relevant to a common event as it unfolds. The framework for this platform is emerging from a collection of data, maps, and applications available for sharing and download from the ArcGIS Resources Local Government—Public Safety Community website to a fully distributed environment where common users interface concurrently across the same space and time.

Virtual USA

In an attempt to overcome the considerable collaboration challenges facing the emergency preparedness and response community, the DHS Science and Technology Directorate, in partnership with state and local agencies, initiated "an end-user driven and federally supported initiative focusing on cross-jurisdictional information sharing among the homeland security and emergency management community."[16] The DHS describes Virtual USA (vUSA) as "a technical system and [set of] operational guidelines to share incident response information through existing systems and geospatial platforms in partnership with local, tribal, state, and federal officials as well as the vendor community. Through Virtual USA, homeland security and emergency management stakeholders will have the capability to quickly access critical information from relevant sources and customize the [map] display of that information based upon the end-user's unique needs to save lives, protect property, and realize operational efficiencies through improved situational awareness."[17]

Virtual USA conducted two pilot projects that brought together regional clusters of local, tribal, state, and federal agencies that house critical homeland security–related information within their respective enterprises. The Southeast Regional Operations Platform Pilot, Phases I and II (SE ROPP II) is composed of ten southeastern states: Alabama, Florida, Georgia, Louisiana, Mississippi, North Carolina (observer), South Carolina, Tennessee (observer), Texas, and Virginia. The Pacific Northwest Pilot includes collaboration between four states: Idaho, Montana, Oregon, and Washington. Each of these pilots concluded with a demonstration in which the vUSA capability was tested during a mock exercise. Various federal agencies also participate in the program,[18] and an additional pilot project involving eight states in the Northeast United States was launched

---

16 "Virtual USA," US Department of Homeland Security, accessed September 5, 2011, http://www.firstresponder.gov/Pages/VirtualUSA.aspx.

17 Ibid.

18 Federal agencies include the DHS Science & Technology Directorate; FEMA's National Response Coordination Center (NRCC), National Exercise Simulation Center (NESC), and Incident Management Systems Integration Division (IMSID); FEMA Region IV; the Emergency Services Sector (ESS); the Sector Coordinating Council (SCC); and the State, Local, Tribal, and Territorial Government Coordinating Council (SLTTGCC).

early in 2012. In order to facilitate the use of vUSA, the DHS, working with its state and local partners, developed a National Memorandum of Understanding that provides policies and procedures governing participation in vUSA.

The interoperability of the vUSA initiative was most recently tested during the National Level Exercise (NLE 2011) in May 2011. The purpose of the exercise was to prepare and coordinate a multijurisdictional integrated response to a national catastrophic event, in this case a major earthquake simulated in the central United States region of the New Madrid Seismic Zone (NMSZ).[19] The Central United States Earthquake Consortium (CUSEC) group of states along the New Madrid seismic zone used vUSA during the exercise as a secured private online cloud for all partner states to share GIS data across jurisdictions. Each state registered seven operational GIS layer services in the vUSA cloud and shared passwords with partner states so each state's emergency management agency could grab access to other states' live data as needed. The result was an integrated common operating platform of shared geospatial data that each agency could view within its own map viewer or the COP map viewer designed for central command.

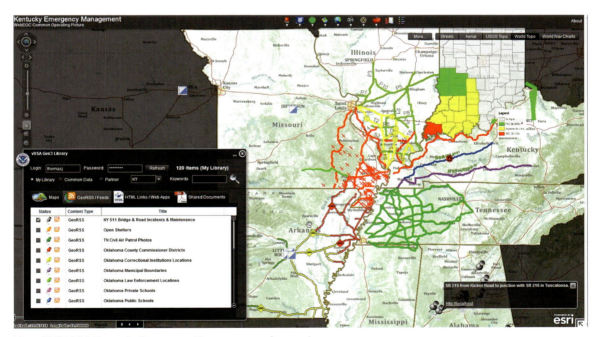

Virtual USA NLE Kentucky Emergency Management web map viewer

Courtesy of G&H International Services, Inc., and Kentucky Emergency Management.

19 "National Level Exercise 2011 (NLE 2011): FACT SHEET," US Department of Homeland Security, Federal Emergency Management Agency, http://www.fema.gov/media/fact_sheets/nle2011_fs.shtm.

The strength of this program is that it integrates existing platforms, visualization tools, and other datasets to allow participating states' systems to interoperate and share information, regardless of the technology. The situational awareness attained by this level of data sharing and distribution is critical to ensuring the safety and security of people and places affected by major catastrophic events that cross boundaries and involve the coordination of multijurisdictional agencies.

Summary

The LGIM provides a fully integrated GIS framework for the daily management of all levels of civic operations, with additional support for public safety planning and emergency operations. The fixed locations of building footprints, facility locations, street centerlines, railroad networks, water bodies, and vegetation are stored as feature classes in the Reference Data feature dataset. These reference features are complemented by additional datasets that store the locations of feature classes such as administrative boundaries and cadastral references, parcels, utility infrastructure, demographic information, land-use planning, and tax assessment information. These fixed reference features form the basemap layers of the LGIM, and the foundation of the Public Safety basemap that adheres to conventional cartographic mapping standards, ensuring interoperability and compatibility with other agency partners active in an event.

Emergency facilities and logistics resource inventories are stored as feature classes in the LGIM in the Public Safety Planning feature dataset and can be called upon during planning and preparedness exercises that depend on the location of these assets. As an event unfolds, the Emergency Operations feature dataset of the LGIM is elevated from observation to activation mode, allowing emergency managers to use incident data feeds to identify the location and impacts of a significant event. Response personnel and resources are then assigned to secure public safety. These dynamic data features form the operational layers of the LGIM.

Additional data that extends the LGIM geodatabase, and which is effectively managed within the ArcGIS platform, include extensive libraries of high-resolution aerial imagery that support emergency planning and operations; dynamic links to documents that outline standard operating procedures and protocols in the event of an incident, supported by digital copies of floor plans and evacuation maps; live data feeds from various national and regional agencies that provide up-to-date traffic, weather, and shelter information critical to duty officers and first responders; VGI made available via web-based digital social media platforms; and data-sharing and distribution initiatives, such as

ArcGIS Online and Virtual USA, which enable interoperability of systems and data during events that cross multijurisdictional boundaries.

All combined, these data and information schema provide a comprehensive foundation from which to ensure the greatest situational awareness before, during, and after a significant event occurs. This foundation is wholly supported by the ArcGIS common operating platform that provides the structural and systematic base to support a fully integrated GIS for emergency management.

Once the GIS is populated with a comprehensive geodatabase of local features, and is supported by additional imagery and live data characterizing a community or an event, it can begin to do what it does best—analyze the data to inform decisions that secure the greatest safety for the greatest number of people. The next chapter explores how traditional and advanced GIS analytical tools are applied to this data to enhance situational awareness of a community or an event as it unfolds.

Chapter 3: Planning and analysis

Planning and analysis is the foundation of emergency operations. Having all of the relevant data embedded in an integrated geodatabase is only one piece of the complete emergency management GIS common operating platform. How to analyze the data to assess a community's vulnerabilities, how best to mitigate those vulnerabilities, where contingency plans need to be built, and what training needs to be provided are another piece. Budget preparation also benefits from planning and analysis as a way to justify funding needs in support of mitigation measures.

Part of planning and analyzing for emergency operations is mitigation planning. The primary goal of the US Disaster Mitigation Act of 2000 is for state and local governments to be able to "implement effective hazard mitigation measures that reduce the potential damage from natural disasters."[1] This act requires communities to complete mitigation plans in order to receive hazard-mitigation funds. GIS supports this effort by modeling how complex natural and technological systems affect critical infrastructure and populations. It employs key analytic tools to use the basemap and operational data layers in the geodatabase to geographically evaluate and determine the potential effect of an event. By creating new data that empowers the geographic knowledge base of a community, emergency managers can quickly determine who is affected by an event and how to minimize the risk. For example, knowing the proximity of landslide-prone hillsides or rivers susceptible

---

1 "US Disaster Mitigation Act of 2000," US Department of Homeland Security, Federal Emergency Management Agency, Resource Record Details, July 2002, http://www.fema.gov/library/view Record.do?id=1935.

to flash flooding is the first piece of information needed to understand how they may influence nearby power stations or elderly residents.

GIS provides extensive analytical tools to researchers studying the many effects and interactions of physical and social systems around the world. In the past, however, these tools and research findings were beyond the reach of emergency personnel with little or no GIS training. GIS analysis capabilities were generally limited to the desktop PC, managed and implemented by skilled GIS analysts within an agency or across an enterprise. Today, the line between desktop GIS analysis capabilities and web-enabled services has blurred. With the development of the ArcGIS for Server common operating platform, geoprocessing tools and services are now available through the web-based situational awareness map viewer, enabling emergency managers and operations personnel to leverage the power and information of the GIS without requiring advanced GIS training or advanced analytical skills.

This extended GIS platform and enhanced capability has also blurred the line between those tools and analysis models available to emergency managers throughout the phases of the emergency management workflow. Data-intensive expert tools that required advanced geoprocessing algorithms and extended amounts of processing time were once only practical in the planning and mitigation stages of emergency management. Streamlined and more efficient adaptations of these tools are now widely available to emergency operations personnel, and to the public at large, through web-enabled map viewers loaded with embedded geoprocessing widgets and geodatabases. Comprehensive vulnerability analysis and consequence modeling can now effectively be integrated into the emergency management workflow before, during, and after an event unfolds. In a rapid response scenario, emergency operations personnel can access a set of web-enabled tools that take risk and vulnerability into account to facilitate a quick determination of who and what is affected by an event and determine how to get public safety resources where they need to be as quickly as possible. Emergency managers are now empowered to reference such actionable information when planning for future events or possible hazards and also when making decisions in the midst of a response. Having ready access to such information at critical times in the life cycle of a significant event greatly enhances the situational awareness of an emergency operations team as it works to return a community to safety.

Analysis for preparedness, mitigation, and response
Knowing what places in a community are at the greatest risk of threat if a catastrophic event occurs is one of the most important pieces of information an emergency manager can have. As plans are formulated to protect

people and property, mitigation measures can be put in place to minimize the known risk and threat. Using GIS to perform a pre-event vulnerability analysis assists planners in this task as they consider how to lessen an expected impact and decide where resources and responders would be best deployed when an event does occur. Integrating the findings of this analysis into the geodatabase as public safety operational layers empowers the common operating platform to provide the greatest level of situational awareness in all phases of an event. Web-based analysis tools and services can also now generate risk and impact studies on the fly to enable quick real-time analysis as an event unfolds and response and recovery are underway.

Vulnerability analysis

The basis for determining the vulnerability of people and property to catastrophic risks in a community is the proximity they have to natural and technological hazards. Because the foundation of GIS is the location of place-based features, it is the most effective tool to use to measure the distance and interactions between them. The analysis starts with knowing where the features are that pose a risk to a community. A multihazard approach enables planners to foresee where those places are that are likely to suffer the greatest damage should catastrophic events occur. Planners can study the basemap and operational layers of the geodatabase to determine the location of technological hazards (such as designated hazmat routes, toxic materials sites, and railroad lines that carry hazardous materials) and areas prone to natural hazards (such as floods, landslides, and earthquakes). By designating buffer zones at determined distances around hazardous sites, planners can determine areas of high, medium, and low risk based on proximity to these areas.[2]

Although the objective is to combine all these risk features (see maps on pages 64–65) into one map to determine the places where they coexist, doing so often creates a complex and confusing visualization (see page 66, top).

Moving beyond the visualization, planners can analyze this information to more accurately determine where these hazardous features coexist. GIS can run a vulnerability analysis model that weighs these features by risk and merges them together into one feature that represents the geographic coexistence of these combined hazardous conditions.

---

2 For a complete account of the vulnerability assessment model, see Eric Tate, Susan L. Cutter, and Melissa Berry, "Integrated Multihazard Mapping," *Environment and Planning B, Environment and Design* 37, no. 4 (2010): 646–63.

Designated hazmat routes with distance buffers Map by Esri; data courtesy of LOJIC, Louisville, Kentucky.

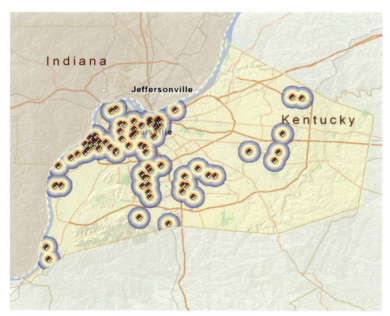

Toxic materials sites with distance buffers Map by Esri; data courtesy of LOJIC, Louisville, Kentucky.

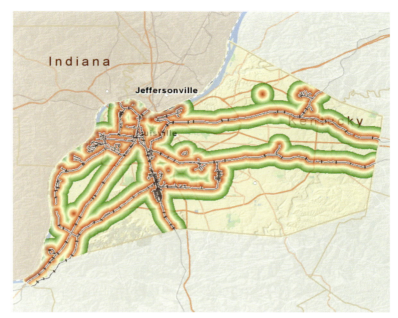

Railroads with distance buffers Map by Esri; data courtesy of LOJIC, Louisville, Kentucky.

100- and 500-year floodplains layer Map by Esri; data courtesy of LOJIC, Louisville, Kentucky.

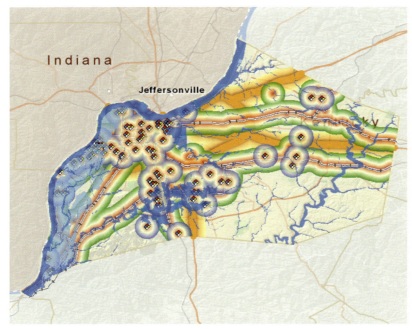

Combined distance buffers and floodplains Map by Esri; data courtesy of LOJIC, Louisville, Kentucky.

Input Hazards

Flooding

Railroads

HAZMAT Facilities

HAZMAT Routes

Risk Weighted Overlay

Hazard Vulnerability

Hazard

Hazard vulnerability layer Map by Esri; data courtesy of LOJIC, Louisville, Kentucky.

In the "Hazard vulnerability layer" figure above, all features are weighted equally (25 percent) because they all pose the same degree of risk to the community. The resulting map shows areas in darkest red where all four hazardous features coexist in the same place. Also included are the areas surrounding these features, shown in graduated shades of color that get lighter with reduced proximity. Areas outside of these shaded regions do not contain hazardous features and are outside the proximity buffer of these known hazardous conditions. This *hazard* feature becomes one of three map layers in the vulnerability analysis model.

The second map layer in the model is the location of critical infrastructure, defined by DHS as "the assets, systems, and networks, whether physical or virtual, so vital to the United States that their incapacitation or destruction would have a debilitating effect on security, national economic security, public health or safety, or any combination thereof."[3] These include community assets such as transportation systems, police and fire facilities, schools, power stations, water resources, commercial facilities, and health-care services. These features are included in the basemap Reference layers of the geodatabase and are ready to use in the planning and analysis of potential events in a community. GIS can map these locations and perform a density analysis to determine choke points where these critical features are most present.

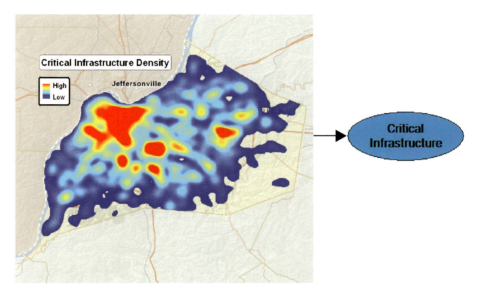

Critical infrastructure density layer Map by Esri; data courtesy of LOJIC, Louisville, Kentucky.

The third and final map layer in the model is the location of at-risk populations in the community (see next page, top). Studying demographic data available from the US Census Bureau[4] reveals places where concentrations of population groups live, such as elderly residents and children younger than age five, as well as the age and value of structures in a neighborhood and where mobile home communities exist.

3 "Critical Infrastructure," US Department of Homeland Security, Federal Emergency Management Agency, http://www.dhs.gov/files/programs/gc_1189168948944.shtm.
4 US Census Bureau, American FactFinder, http://factfinder2.census.gov.

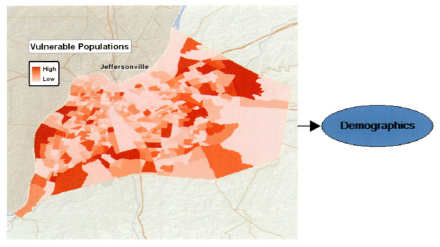

Vulnerable populations layer Map by Esri; data courtesy of LOJIC, Louisville, Kentucky.

These variables show the places where emergency operations planners need to concentrate special services for evacuation and response to secure the safety of these at-risk populations should a hazardous event occur nearby.

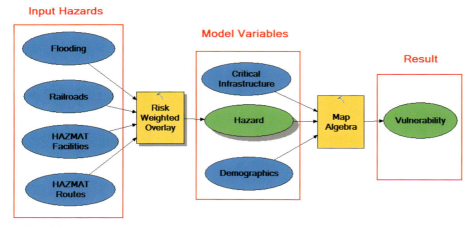

Vulnerability Assessment model Esri.

The final model defines the GIS workflow that identifies those areas most vulnerable to disasters. The analysis performs *map algebra* by applying a derived risk factor to the weighted input hazards to arrive at a social vulnerability index for each location in the community.[5]

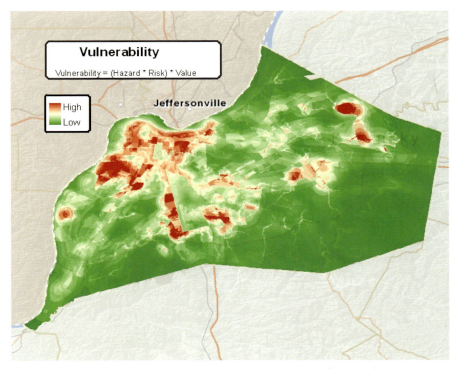

County vulnerability layer Map by Esri; data courtesy of LOJIC, Louisville, Kentucky.

The overall output shows those areas in red where a hazardous event is most likely to have the greatest effect on a community. This map can be included in an overall mitigation plan document for an agency or organization. It also becomes an important map layer in the GIS geodatabase used by emergency management planners to make decisions about mitigation priorities aimed at reducing the chances that

---

5 The methodology for this analysis is based on the work done by S. L. Cutter, J. T. Mitchell, and M. S. Scott in "Revealing the Vulnerability of People and Places: A Case Study of Georgetown County, South Carolina." *Annals of the Association of American Geographers* 90, no. 4 (2000): 713–37. A simplified version of this index has been compiled for the entire United States, summing eight variables to arrive at a surrogate for social vulnerability, the SoVI index. This map is available as a web service on the "USA Social Vulnerability" page of the ArcGIS.com website, http://www. arcgis.com/home/item.html?id=b5501cc71fe44f8d9a0df362ea6aebb3. A more inclusive analysis taking into account a full set of forty-two variables has been undertaken by the Hazards and Vulnerability Research Institute at the University of South Carolina. Details of this study can be reviewed at the Hazards and Vulnerability Research Institute website, http://webra.cas.sc.edu/hvri/ products/sovi_details.aspx.

such an event could occur, and, if it did, to lessen its effect on the surrounding community. Such measures might include reassignment of hazmat routes away from at-risk populations during particular times of day, pre-event designation of assembly areas outside the affected zone in case of evacuation, and the stockpiling of food and medical supplies at these locations for easy deployment during response.

Specific vulnerability assessments can also be performed to model the potential risk of singular catastrophic events. Narrowing the input variables in the model to proximity to fault lines to study the potential risk of seismic activity in a region, or nearness to the urban–wildland interface to model wildfire risk, are effective ways of adapting this model to gauge the vulnerability that communities may have to specific catastrophic events that may threaten their safety.

Web-enabled vulnerability assessment tools

The results of a GIS-based vulnerability assessment of a community can be effectively integrated into a web-enabled situational awareness viewer for quick and easy access by all partners in an emergency management operation. Public viewers are also available to citizens in a community to gauge their vulnerability to potential hazards found in their neighborhoods.

The Integrated Hazard Assessment Tool

At the time of this writing, the Hazards and Vulnerability Research Institute (HVRI) at the University of South Carolina is developing a "portal for web-based hazard and social vulnerability assessments. The primary purpose of this application is to visualize and map hazard frequencies, social vulnerability, and event information to enable informed vulnerability assessments that conform to the Disaster Management Act (DMA) 2000 standards."[6] The Integrated Hazard Assessment Tool (IHAT) is formulated on the model discussed earlier that determines social vulnerability to environmental hazard events. Additional analysis by the HVRI intersects hazard vulnerability and social vulnerability to yield a "place vulnerability" geography at the subcounty level. The IHAT application lets users select a county in South Carolina, turn on a select set of critical facilities features, and view the area within the context of the social, biophysical, and place vulnerabilities. A linked data page provides a sortable and downloadable tabular view of the number of loss-causing events, hazard frequencies, and annualized losses for any county in the state for the time frame 1960–present.

---

6 "Integrated Hazard Assessment Tool (IHAT), Beta Version," University of South Carolina, Arts and Sciences, Hazards and Vulnerability Research Institute, http://webra.cas.sc.edu/ihat/.

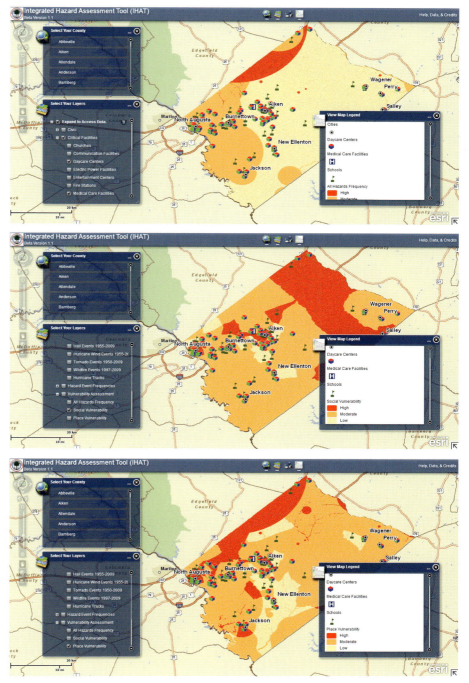

IHAT: All hazards, social vulnerability, and place vulnerability Courtesy of Hazards and Vulnerability Research Institute, University of South Carolina.

The value that this application lends to emergency management and planning is in its time savings and ability to quickly identify "areas of greatest need" from a decision-making perspective. Users need not know the ins and outs of GIS analysis/programming or desktop GIS to leverage this application. In fact, the four-click process to create a map for use in planning, reporting, or decision making lets users spend less time in the process of gathering data and creating maps and more time analyzing patterns and preparing efficient mitigation and response tactics. Additionally, the application provides a means for users to quickly focus on those areas where populations may face the greatest risk of disaster events and suffer the greatest impact from those events because of their inherent social vulnerability. Planners can use this tool to direct pre-event resources to those areas with higher vulnerability to disasters, mitigating the negative effects that future events may impose on selected communities, or they can pre-position rapid response resources during the run-up to a major event based on the identification of at-risk and vulnerable populations.

Multi-Hazard Risk Tool
The Buncombe County, North Carolina, EOC, in collaboration with the RENCI Engagement Center at University of North Carolina at Asheville and the RENCI headquarters at Chapel Hill, has developed the GIS-based Multi-Hazard Risk Tool that is available online to county planners, emergency management personnel at the EOC, and the public at large. A countywide vulnerability analysis has been performed on a desktop GIS to identify the level of risk faced by the threat of floods, wildfires, and landslides in the region. Each scenario is prepared in the map viewer as a separate theme to enable an assessment of how the effects of each of these potential hazards might play out when an event takes place.

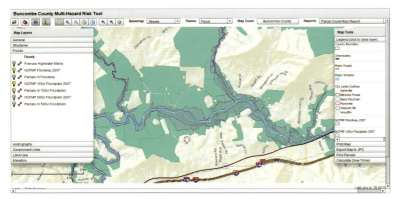

Buncombe County, North Carolina, Multi-Hazard Risk Tool Courtesy of National Environmental Modeling and Analysis Center (NEMAC), University of North Carolina at Asheville.

The Multi-Hazard Risk Tool flood theme map (see page 72) enables users to see which parcels in the county are within the 100-year floodplain.

Parcel Count for Floods for Buncombe County   ×

DRAFT - Do not use numbers for any official capacity.   `Excel Export`

| Floodway | 100 Year Floodplain | 500 Year Floodplain | Information |

Agricultural	Parcels	% Total Parcels	Acres	% Total Acres
Occupied	74	0.05	7,978.58	2.00
Vacant	40	0.02	1,959.20	0.49
Total	114	0.09	9,937.79	2.50
Commercial	**Parcels**	**% Total Parcels**	**Acres**	**% Total Acres**
Occupied	407	0.34	5,049.41	1.26
Vacant	145	0.11	548.80	0.14
Total	552	0.46	5,598.22	1.40
Industrial	**Parcels**	**% Total Parcels**	**Acres**	**% Total Acres**
Occupied	33	0.02	2,162.42	0.54
Vacant	0	0.00	0.00	0.00
Total	33	0.02	2,162.42	0.54
Residential	**Parcels**	**% Total Parcels**	**Acres**	**% Total Acres**
Occupied	979	0.81	3,478.56	0.87
Vacant	315	0.25	737.35	0.18
Total	1,294	1.07	4,215.91	1.05
Other	**Parcels**	**% Total Parcels**	**Acres**	**% Total Acres**
Occupied	178	0.15	7,947.00	2.00
Vacant	177	0.15	2,158.33	0.54
Total	355	0.28	10,105.33	2.53
Total	**Parcels**	**% Total Parcels**	**Acres**	**% Total Acres**

Parcel report Courtesy of National Environmental Modeling and Analysis Center (NEMAC), University of North Carolina at Asheville.

Parcel count and value reports can also be generated and exported to a spreadsheet to assess the specific effects of a possible event. Similar map themes and reports are included to address the potential effects of wildfires and landslides. The tool also lets the user add other map layers to the viewer, such as schools or day care centers, to see the proximity of vulnerable populations to high-risk areas.

An additional map theme (see page 74) combines the three potential hazards into an all-hazards scenario to enable a comprehensive assessment of the most vulnerable areas at risk in the county. Planners are able to view the potential risks of these hazards concurrently by overlaying various map layers to determine where their greatest resources should go, where to design mitigation plans to minimize the extent of damage if an event occurs, and how to ultimately secure the safety of its citizens and the protection of its property.

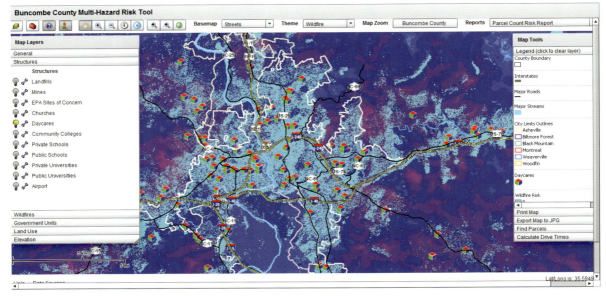

Additional map layers added to map viewer Courtesy of National Environmental Modeling and Analysis Center (NEMAC), University of North Carolina at Asheville.

Texas Wildfire Risk Assessment Portal

Another example of a web-enabled risk-assessment tool is the Texas Wildfire Risk Assessment Portal (TxWRAP). Texas Forest Service, in partnership with DTS (Data Transfer Solutions) (Orlando, Florida), developed this portal to deliver the results of an extensive statewide risk assessment providing a "consistent, comparable set of scientific results to be used as a foundation for wildfire mitigation and prevention planning in Texas."[7] The application, built on an ArcGIS for Server operational platform, provides access to a suite of geoweb applications, including the following:

- *Public viewer*—Enables users in the general public to assess their own risk by identifying "specific risk levels within a 2-mile radius of a home, or any other point of interest on the map, and provides a link to additional resources for users wanting to know how to reduce their risk."[8]
- *Professional viewer*—"Designed to support the community wildfire protection planning needs of government officials, hazard-mitigation planners and wildland fire professionals. This application contains advanced functionality and additional map themes as compared to the public viewer. The key features of this application include the capability to define a project area, generate a detailed risk summary

7 "Texas Wildfire Risk Assessment Portal: Are You at Risk?" Texas Forest Service, 2010, http://www.texaswildfirerisk.com/Home/Risk/.

8 Ibid.

report, and export and download wildfire risk GIS data. Tools also integrate Esri's Business Analyst Online data services to generate on-the-fly impact analysis for planning and operational requirements."[9]

◆ *Fire Occurrence Explorer*—"Designed to analyze historical wildfire occurrence data. The featured tool in this application is called the 'Fire Cause Analyzer.' It allows users to interactively visualize and explore statewide or county fire reporting data using dynamic charting and mapping capabilities in order to derive patterns related to fire cause. This application helps local planners analyze fire causes so they can better evaluate prevention program options."[10]

◆ *Community Assessment Editor*—"A web-mapping application that allows approved users to create and manage wildfire assessments at the community level. This includes community, home, and post-fire home assessments. All information is collected using standardized forms and is stored in a centralized database."[11] This application integrates with a geomobile application for data collection and seamless data synchronization, providing a foundation for integrating field-collected data with risk-assessment data.

The results of an assessment can be used to help prioritize areas in the state where tactical analyses, community interaction and education, or mitigation treatments may be necessary to reduce risk from wildfires.

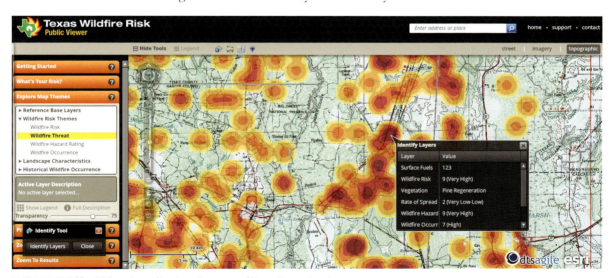

TxWRAP Wildfire Risk web application Courtesy of Texas Forest Service.

9 "Texas Wildfire Risk Assessment Portal: Are You at Risk?" Texas Forest Service, 2010, http://www.texaswildfirerisk.com/Home/Risk/.
10 Ibid.
11 Ibid.

Modeling consequences of specific hazard events

Although a vulnerability analysis determines those areas where a catastrophic event may have the greatest effect, it does not always define how the hazard itself may unfold in a community. Consequence models are designed to predict how certain hazards can potentially behave and what, where, and how the damage may occur. For example, scenarios can be configured to model how a wildfire behaves in certain regions given a set of environmental conditions that affect its rate of spread. Models can be applied in the planning stages of emergency management to estimate the effects of potential events, or they can be deployed as an event unfolds to assist operations personnel as they respond to an incident that threatens the immediate safety of a community.

Tools have been developed to move these models out from the laboratory and into the hands of planners and responders as they face the challenges of protecting communities. Such tools range from desktop applications that process extensive databases of physical environmental features and detailed inventories of the built infrastructure to web-based GIS widgets that leverage the findings of these models into quick and effective hands-on tools for real-time response and recovery operations. The following review is by no means exhaustive, but rather offers a sample of the types of tools available to assist emergency management personnel as they integrate GIS into their mitigation planning and response operations.

Hazardous materials

Returning to the case of the overturned tanker, a number of GIS-enabled tools are available to assist emergency planners and operations personnel manage and control the effects that such an event may have on a the safety of a community.

Chemical incident

In the moments just after a significant event occurs, such as the hazardous material released from the overturned tanker, emergency managers must determine the immediate effects of the event. Without any pre-event planning, the emergency manager immediately has access to a set of analytical GIS tools present in the ERG toolkit. This toolkit is already embedded in the situational awareness map viewer template available from the ArcGIS Resources Local Government—Public Safety Community website. It is readily available to users without any additional programming, enabling the quick and effective analysis of the basemap data and operational layers of the geodatabase within the first 30 minutes of an event.

The parameters embedded in the tool come directly from the *Emergency Response Guidebook 2008* (ERG2008), which is used by firefighters, police, and other emergency services personnel who arrive at the scene of a transportation incident involving a hazardous material.[12]

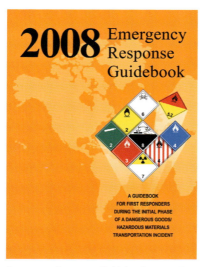

Emergency Response Guidebook 2008 Courtesy of US Department of Transportation, Pipeline and Hazardous Materials Safety Administration.

The guidebook was developed jointly by the US Department of Transportation, Transport Canada, and the Secretariat of Communications and Transportation of Mexico (SCT). The guidebook itself inventories hundreds of hazardous chemicals. For each chemical, recommended safe distances are listed, including "initial isolation distances," within which all persons should be considered for evacuation in all directions from the actual spill or leak source, and "protective action distances," which are areas in which first responders should evacuate people to protect safety.[13]

The GIS-based ERG tools place all input parameters that the model requires to calculate the designated zones onto the map as polygons representing those areas.

---

12 More information on the ERG can be found at the US Department of Transportation and Hazardous Materials Safety Administration website: http://phmsa.dot.gov/hazmat/library/erg. A PDF of the *Emergency Response Guidebook 2008* can be found at http://phmsa.dot.gov/staticfiles/PHMSA/DownloadableFiles/Files/erg2008_eng.pdf.

13 Jeff Baranyi, "Emergency Response Guide Geoprocessing Tools for ArcGIS," *ArcGIS Resource Center/Public Safety* (blog), June 15, 2009, http://blogs.esri.com/Dev/blogs/publicsafety/archive/2009/06/15/Emergency-Response-Guide-Geoprocessing-tools-for-ArcGIS.aspx.

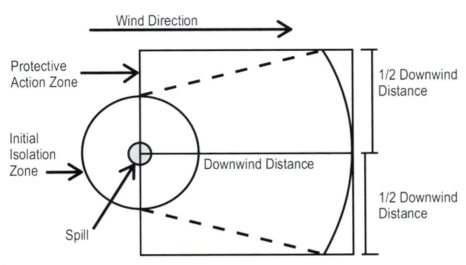

Wind Direction

Protective Action Zone

1/2 Downwind Distance

Initial Isolation Zone

Downwind Distance

1/2 Downwind Distance

Spill

ERG safe distances Esri.

In the case of the overturned tanker, the first thing response personnel need to do is to determine the affected area. Prior to this incident, the city implemented a well-designed GIS data management strategy embedded in an ArcGIS common operating platform, as described in chapter 2. The duty officer is now able to leverage the base and operational layers of that comprehensive geodatabase to deploy the ERG tools embedded in the situational awareness viewer to quickly model the area of the plume or spill.

ERG data input screen
Esri.

By entering the material type as reported by the police officer on site and communicated via the live 9-1-1 incident data feed, and the wind direction, as reported via a live feed from the closest weather station, the duty officer can click the point on the map where the derailment occurred to quickly characterize the effect of the spill without overcomplicating the display.

The resulting map shows the initial isolation zone in pink and the protective isolation zone in red (see page 80, top). Emergency personnel can now identify affected critical infrastructure and calculate the number of people living within the affected area.

The GIS can then generate a list of street addresses within the designated zones to be fed into the mass notification system to alert people in these areas of impending lockdown, shelter-in-place, or evacuation orders (see page 80, bottom). The ERG tool also provides emergency personnel with demographic data about the affected residents to inform decisions about evacuation and shelter needs.

In this case, a large population of school-age children in the area will require family shelter and access to continued school activities (see page 81, top).

The immediate use of this ERG tool creates powerful actionable information quickly created by the GIS to address the urgent needs of the community within the early stages of the incident. By implementing a sound data management strategy embedded in an ArcGIS common operating platform, emergency operations personnel can effectively leverage the data stored and managed in the system to quickly provide actionable information to ensure the safety of the community.

Moving forward beyond the initial 30 minutes of a hazmat incident, emergency managers can draw from additional planning and analysis tools to support the response and recovery stages of such an event. CAMEO (Computer-Aided Management of Emergency Operations)[14] is a set of applications designed to model the effects of a chemical incident. It integrates a database of more than six thousand hazardous chemicals with a chemical reactivity function that predicts what happens when chemical substances are mixed together. Also included is MARPLOT (Mapping Applications for Response, Planning, and Local Operational Tasks), a basic mapping module that uses US Census map files to display areas of potential impact, and ALOHA, an air-dispersal model that generates a plume layer that estimates the location of threat zones in an affected area. This plume layer can be effectively integrated into the situational awareness viewer, using the ALOHA ArcMap Import Tool from NOAA[15] with the results of the ERG to map out a more precise impact zone resulting from a chemical spill. Response personnel can then be quickly deployed to set up road closures and access points along the perimeter of the zone (see page 81, bottom).

14 "Computer-Aided Management of Emergency Operations (CAMEO)," US Environmental Protection Agency/Emergency Management, http://www.epa.gov/oem/content/cameo/.

15 "ALOHA ArcMap Import Tools," Office of Response and Restoration/NOAA's Ocean Service/ National Oceanic and Atmospheric Administration/US Department of Commerce, http://archive. orr.noaa.gov/topic_subtopic_entry.php?RECORD_KEY(entry_subtopic_topic)=entry_id,subtopic_ id,topic_id&entry_id(entry_subtopic_topic)=528&subtopic_id(entry_subtopic_topic)=24&topic_ id(entry_subtopic_topic)=1.

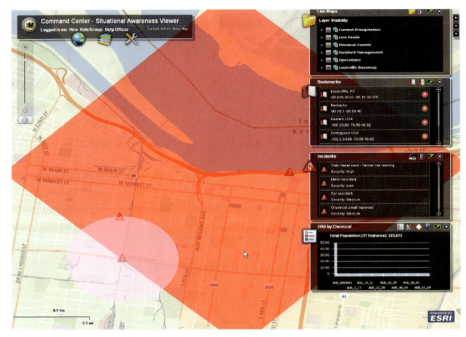

ERG isolation zones Map by Esri; data courtesy of LOJIC, Louisville, Kentucky.

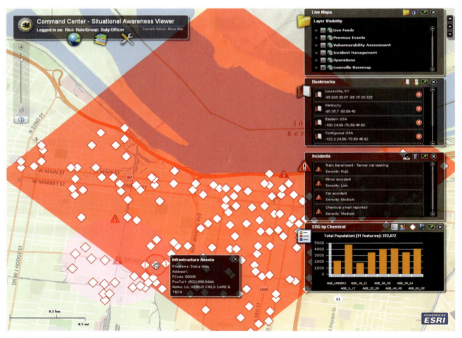

Affected facilities in isolation zones Map by Esri; data courtesy of LOJIC, Louisville, Kentucky.

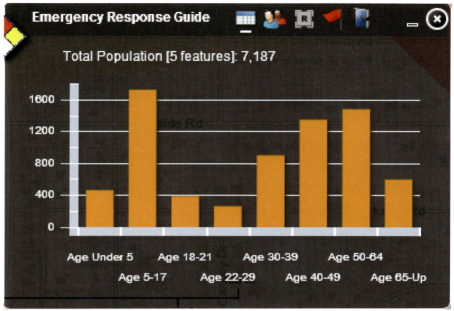

Affected population in isolation zones Esri.

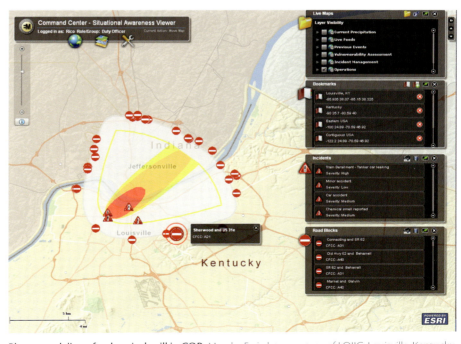

Plume modeling of a chemical spill in COP Map by Esri; data courtesy of LOJIC, Louisville, Kentucky.

Additional analysis tools readily available within the situational awareness viewer include the Find Closest Resource, successfully employed in the Kansas-Map v2 application. This tool generates a list of the closest mutual aid FEMA and local emergency resources queried, and then shows the route on the map, along with driving directions detailing the quickest way to get to the resource.

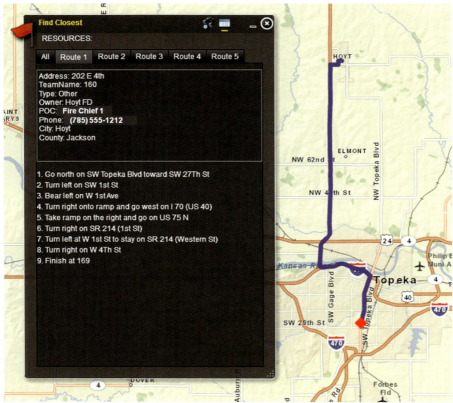

Kansas-Map v2 Find Closest Resource tool Courtesy of Kansas Adjutant General's Department, Division of Emergency Management.

The Drive Time analysis tool (see next page) is an additional widget that generates driving-distance bands around a designated location based on drive-time distances between 15 and 40 minutes.

These straightforward analysis tools are readily available in the situational awareness viewer and require little or no programming or setup parameters other than those already configured in the base and operational layers of the geodatabase residing in the ArcGIS common operating platform. They are easily deployed in both planning and

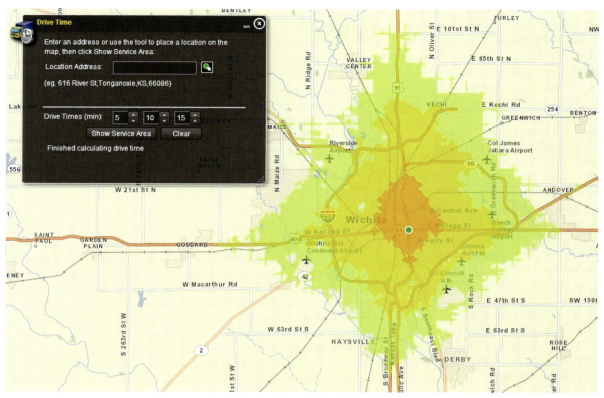

Kansas-Map v2 Drive Time tool Courtesy of Kansas Adjutant General's Department,
Division of Emergency Management.

operations settings to assist emergency management personnel in mitigating the potential impacts of an expected or unfolding hazardous event.

Oil spill

More modeling tools are available to map the path of a hazardous event occurring from a potential or current oil spill. NOAA's GNOME (General NOAA Oil Modeling Environment)[16] spill response trajectory model analyzes different pollutants and environmental conditions in varying scenarios to predict how wind, currents, and other processes might move and spread oil spilled on or under the water surface. These results are coupled with other response models, such as ADIOS2 (Automated Data Inquiry for

16 "GNOME," Office of Response and Restoration/NOAA's Ocean Service/National Oceanic and Atmospheric Administration/US Department of Commerce, http://archive.orr.noaa.gov/type_subtopic_entry.php?RECORD_KEY%28entry_subtopic_type%29=entry_id,subtopic_id,type_id&entry_id%28entry_subtopic_type%29=292&subtopic_id%28entry_subtopic_type%29=8&type_id%28entry_subtopic_type%29=3.

Oil Spills),[17] which integrates a library of approximately one thousand oils with a short-term oil weathering and cleanup model, and CDOG (Comprehensive Deepwater Oil and Gas Blowout Model),[18] which analyzes the chemical behavior of the oil to determine how and when the oil will reach the surface.

The trajectory results are fully integrated into the ArcGIS for Desktop environment via an import tool that lets planners and responders map the expected course of a spill, along with other map layers that bring vulnerable populations and infrastructure into the scenario. These maps become critical preparedness and operations documents, letting emergency managers deploy resources effectively to safeguard people and places in the path of a spill. The trajectory layers also become vital data in the geodatabase of the common operating platform shared by all partners in an emergency response operation.

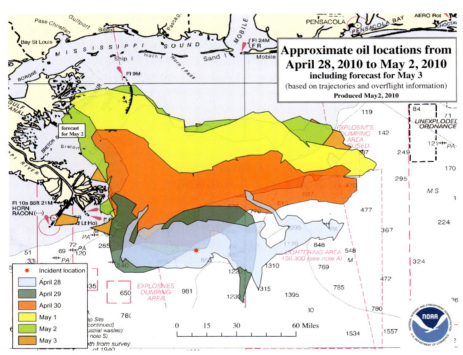

Trajectory path of oil spill generated using NOAA's GNOME analysis tool   Courtesy of NOAA's Office of Response and Restoration.

17 "ADIOS2," Office of Response and Restoration/NOAA's Ocean Service/National Oceanic and Atmospheric Administration/US Department of Commerce, http://archive.orr.noaa.gov/type_topic_entry.php?RECORD_KEY%28entry_topic_type%29=entry_id,topic_id,type_id&entry_id%28entry_topic_type%29=181&topic_id%28entry_topic_type%29=1&type_id%28entry_topic_type%29=3.
18 Poojitha D. Yapa, "Comprehensive Deepwater Oil and Gas Blowout Model," Department of Civil and Environmental Engineering, Clarkson University, http://www.cdogmodel.com/.

In the case of the Deepwater Horizon Gulf oil spill, how the oil would travel once it surfaced was effectively modeled using GNOME and mapped in a GIS to display the expected trajectory of the spill path. This actionable information was "used by the response team to calculate environmental trade-offs for different response options."[19]

HAZUS-MH

HAZUS (Hazards–United States)-MH is a free application developed in 1997 by FEMA to provide earthquake loss estimation methods to state and local governments to assess the potential effects of seismic events that may threaten their communities.[20] It was further enhanced over the past decade to include methods and tools to assess the potential effects of flood and hurricane events across the United States.[21] Embedded in the HAZUS-MH application is a set of engineering and scientifically based methodologies that use an organization's existing ArcGIS geodatabase and operating platform to map and display hazard data and the results of damage and economic loss estimates for buildings, infrastructure, and populations.

Earthquake model

The earthquake model used in HAZUS-MH considers the seismic history, building stock, and local geology of a region to determine how a hypothetical seismic event might affect a community. The output of the model produces ArcGIS maps of potential scenarios showing projected ground shaking and the extent of physical damage, economic loss, and social disorder that activity is expected to cause the local population. The maps are supported by charts and tables that detail the number and types of buildings damaged, number of potential casualties, damage to transportation and utility infrastructure, people displaced from their homes requiring shelter and medical needs, lost jobs and business interruptions, and the estimated cost of repairing the projected damage.[22]

The HAZUS-MH earthquake model has been used extensively in a wide range of applications by federal, state, local, and private sectors users. Studies include the San Francisco Bay Area Earthquake Exercise in 2006, scenario

19 Stephanie Pappas, "Gulf Oil Spill Cleanup Gets Assist from Virtual Reality," *Live Science*, May 4, 2010, http://www.livescience.com/6418-gulf-oil-spill-cleanup-assist-virtual-reality.html.

20 HAZUS can be ordered free of charge from https://msc.fema.gov/webapp/wcs/stores/servlet/CategoryDisplay?catalogId=10001&storeId=10001&categoryId=12013&langId=-1&userType=G&type=14.

21 Philip J. Schneider and Barbara A. Schauer, "HAZUS—Its Development and Its Future," American Society of Civil Engineers, doi: 10.1061/ASCE 1527-6988 2006 7:2 40, http://ascelibrary.org/nho/resource/1/nhrefo/v7/i2/p40_s1?isAuthorized=no.

22 "HAZUS," Federal Emergency Management Agency, http://www.fema.gov/hazus/.

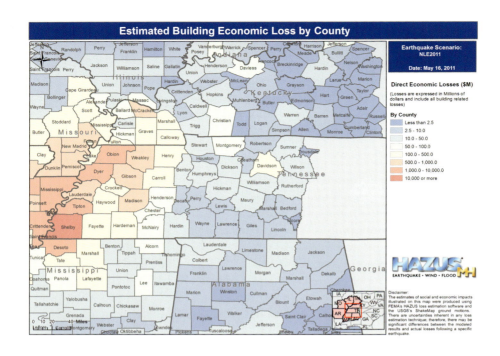

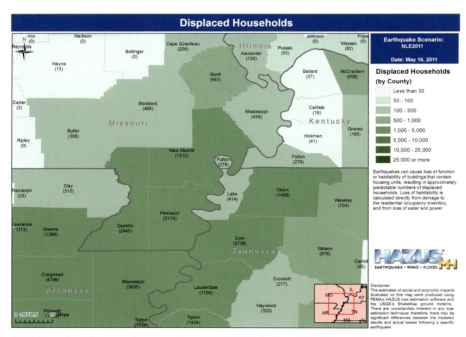

HAZUS earthquake scenario maps Courtesy of FEMA Region VIII Mitigation.

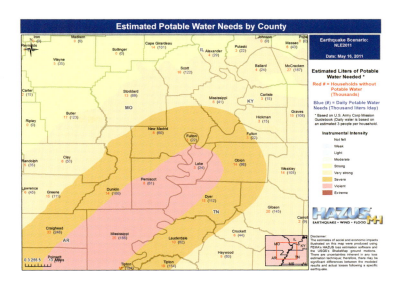

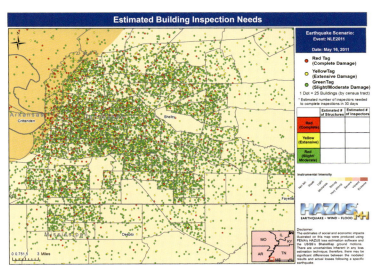

HAZUS earthquake scenario maps Courtesy of FEMA Region VIII Mitigation.

development for the New Madrid Seismic Zone Catastrophic Planning Initiative, and a loss estimation study undertaken by Orange, Riverside, and San Bernardino counties in Southern California to analyze the potential effects of two earthquake scenarios on hospital services in the area.[23]

23 "HAZUS-MH Used to Support San Francisco Bay Area Earthquake Exercise," Federal Emergency Management Agency, http://www.fema.gov/library/viewRecord.do?id=4321; "Application of HAZUS to the New Madrid Earthquake Project," Federal Emergency Management Agency, http://www.fema.gov/library/viewRecord.do?fromSearch=fromsearch&id=4314; Federal Emergency Management Agency, http://www.fema.gov/plan/prevent/hazus/eq_hosp_analysis.shtm.

Estimated Building Inspection Needs

Table 1 depicts the estimated number of building inspectors that would be required in the event of an earthquake in the study region based on teams of 2 inspecting 10 buildings per day. A total estimate of **8,530** inspectors are needed to complete inspections in **30 days** for **1,030,799** buildings. Only counties with at least slight expected buildings damage were included.

Table 1: Estimated Building Inspection Needs

County	# Buildings	# Inspectors
Alabama		
Bibb	168	3
Blount	383	4
Calhoun	205	3
Cherokee	107	2
Chilton	241	3
Clay	25	1
Cleburne	25	1
Colbert	318	4
Coosa	24	1
Cullman	608	6
Dekalb	465	5
Etowah	379	4
Fayette	145	2
Franklin	216	3
Greene	128	2
Hale	199	3
Jackson	416	4
Jefferson	1,734	13
Lamar	1,043	9
Lauderdale	632	6
Lawrence	287	3
Limestone	371	4

Direct Economic Losses For Buildings

May 16, 2011   All values are in thousands of dollars

	Capital Stock Losses					Income Losses				
	Cost Structural Damage	Cost Non-struct. Damage	Cost Contents Damage	Inventory Loss	Loss Ratio %	Relocation Loss	Capital Related Loss	Wages Losses	Rental Income Loss	Total Loss
Alabama										
Bibb	59	274	133	4	0.03	25	7	17	14	533
Blount	157	1,833	1,191	42	0.07	61	21	41	38	3,384
Calhoun	109	944	632	25	0.01	48	23	44	33	1,858
Cherokee	49	249	142	12	0.02	19	8	15	13	507
Chilton	102	510	271	13	0.03	42	17	34	27	1,015
Clay	9	91	62	7	0.01	3	1	4	2	179
Cleburne	8	63	38	2	0.01	2	1	3	2	119
Colbert	221	5,922	4,614	190	0.15	125	58	102	80	11,313
Coosa	7	54	32	2	0.01	2	0	2	1	101
Cullman	325	4,122	2,938	147	0.08	171	66	138	94	8,002
Dekalb	202	1,790	1,153	88	0.05	88	35	78	59	3,494
Etowah	249	1,725	1,072	38	0.03	143	73	138	90	3,529
Fayette	76	1,028	797	34	0.08	52	18	42	21	2,068

HAZUS earthquake data tables Courtesy of FEMA Region VIII Mitigation.

Flood model

The HAZUS-MH flood model performs flood-hazard analysis and flood-loss estimation analysis. The flood-hazard analysis module uses characteristics, such as frequency, discharge, and ground elevation, to estimate potential riverine and coastal flood depth and elevation, and flow velocity. The flood-loss estimation module calculates physical damage to buildings, transportation and utility lifelines, agricultural areas, and vehicles, and estimates the economic loss and social impacts from the results of the hazard analysis.[24]

The flood model has been widely used by state and local officials for risk assessment and mitigation planning, including Louisiana State University's work using HAZUS to study its flood-prone state; the Johnson County Emergency Management Agency, in collaboration with the University of Iowa, to assess the potential effects of an approaching flood event in June of 2008; and Maryland's comprehensive vulnerability assessment of the state's built environment to riverine and coastal flooding.[25]

Hurricane wind model

The HAZUS-MH hurricane wind model employs a state-of-the-art wind-field model, which has been calibrated and validated using full-scale hurricane data. The model incorporates sea-surface temperature in the boundary layer analysis and calculates wind speed as a function of central pressure, translation speed, and surface roughness.[26] It uses a wind hazard-load-damage-loss framework and considers wind pressure, windborne debris, duration/fatigue, and rain in the analysis. Like its companion models, it calculates physical damage to buildings, transportation and utility lifelines, and civic infrastructure, and estimates the economic loss and social impacts of potential wind damage from an anticipated hurricane event. It also includes

---

24 "HAZUS-MH Flood Model," US Department of Homeland Security, Federal Emergency Management Agency. http://www.fema.gov/plan/prevent/hazus/hz_flood.shtm.

25 "Louisiana State University: Using the Power of HAZUS to Study a Flood-Prone State," US Department of Homeland Security, Federal Emergency Management Agency, Michelle L. Barnett, Center for Energy Studies, Louisiana State University, http://www.fema.gov/library/viewRecord.do?id=3320; "HAZUS User Groups Success Story: FEMA Region VII: Johnson County Emergency Management and Homeland Security Collaborates with the University of Iowa Department of Geography to Assess Flooding Impacts Using HAZUS-MH," Shane Hubbard, Department of Geography, The University of Iowa, Sue Evers, HAZUS Program Manager FEMA Region VII, US Department of Homeland Security, Federal Emergency Management Agency, http://www.fema.gov/library/viewRecord.do?id=3383; "An Assessment of Maryland's Vulnerability to Flood Damage," John M. Joyce, Flood Hazard Mitigation Section, Maryland Department of the Environment, and Michael S. Scott, Eastern Shore Regional GIS Cooperative, Salisbury University, August 2005, http://franklin.salisbury.edu/esrgc/pdf/hazus/appendix_a.pdf.

26 "HAZUS-MH Hurricane Wind Model," US Department of Homeland Security, Federal Emergency Management Agency, http://www.fema.gov/protecting-our-communities/hazus-multi-harzard-hurricane-wind-model.

a tree coverage dataset that enables the estimation of expected tree losses in an affected area.

Applications of the hurricane wind model include the deployment of the tool during the onslaught of Hurricane Irene in 2011 to estimate the extreme wind gusts as the storm made landfall across the eastern United States; a Harris County, Texas, study using HAZUS-MH to undertake an extensive hazard risk assessment for the county; and the use of the model by Dare County, North Carolina, to estimate the potential damage to infrastructure and buildings, and resulting economic impacts to the region, during a wind event.[27]

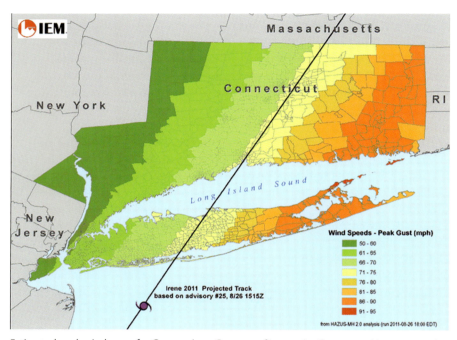

Estimated peak wind gusts for Connecticut Courtesy of Innovative Emergency Management, Inc.

CATS

An additional hazard-modeling application that interfaces with the ArcGIS platform is the Consequences Assessment Tool Set (CATS). This consequence management tool package, available free of charge to US

---

27 Jessica Phillips, Dare County Emergency Management: North Carolina HAZUS User Group, and Eric C. Coughlin, "HAZUS User Groups Success Story: Incorporating HAZUS-MH Methodologies to Assist with Mitigation Planning and Hurricane Operations," http://www.fema.gov/library/viewRecord.do?id=3751, http://www.fema.gov/graphics/plan/prevent/hazus/100yr_floodmap2.gif.

federal, state, and local government emergency response organizations, integrates hazard prediction and containment, consequence assessment, and routing using critical population and infrastructure data.[28] The application focuses primarily on the consequences of technological disasters on population, resources, and infrastructure. Developed in association with the US Defense Threat Reduction Agency (DTRA), CATS is used to address the risks and impacts of such hazards ranging from industrial accidents to acts of terrorism.

CATS tools are integrated into ArcGIS for Desktop and are used to identify roadblocks, populations and infrastructure at risk, atmospheric plumes, recovery resource distribution, blast effects, and road network routing and address locations. Models included within the application include a high-explosive (HE) model; the ERG 2008 tool, as described earlier; NOAA's ALOHA plume model for chemical releases; import of natural hazard damage files from the HAZUS-MH application; DTRA's Hazard Prediction and Assessment Capability (HPAC) model; and the Joint Effects Model (JEM) that predicts and tracks nuclear, biological, chemical (NBC) and toxic industrial chemical/material (TIC/TIM) events and effects.

Levee failure and inundation models

The Resilient and Sustainable Infrastructure Networks (RESIN) project at the University of California, Berkeley,[29] identified Sherman Island, which sits at a maximum of 23 feet below sea level in the Sacramento River Delta, as a major choke point for critical infrastructure. This environment is ideal for modeling the impacts of levee failure resulting from natural hazards, such as earthquakes and storm events. The strategy employs simulated levee failures and island flooding to model and quantify the consequences of accessibility loss for first responders to all citizens living in the inundated region. The simulation floods, measures, and maps the consequences of inundation to road network accessibility. The model uses a "bathtub" flood-simulation approach, filling the entire island from lowest to highest elevations.

Figure A maps the baseline accessibility of first responders in the first 30 minutes. Figure B shows the flooded areas to a height of –14 feet mean sea level, a 10-foot rise in the water level above the lowest elevation point

---

28 "Consequences Assessment Tool Set (CATS)," Science Applications International Corporation (SAIC), 2012, http://www.saic.com/products/security/cats.
29 J. D. Radke, "EFRI-RESIN: Assessing and Managing Cascading Failure Vulnerabilities of Complex, Interdependent, Interactive, Adaptive Human-Based Infrastructure Systems," NSF EFRI-RESIN Workshop on Modeling Sustainable, Resilient, and Robust Infrastructure Systems, University of Illinois, Champaign-Urbana, Illinois (NSF EFRI-0836047), November 17, 2011.

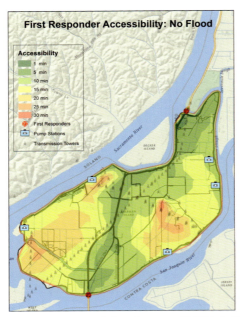

Figure A: Effects of levee failure on accessibility Courtesy of RESIN: Resilient and Sustainable Infrastructure Networks, University of California, Berkeley.

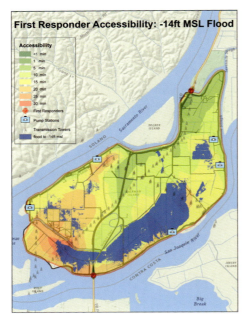

Figure B: Effects of levee failure on accessibility Courtesy of RESIN: Resilient and Sustainable Infrastructure Networks, University of California, Berkeley.

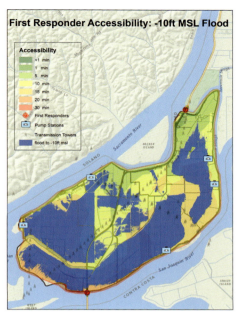

Figure C: Effects of levee failure on accessibility Courtesy of RESIN: Resilient and Sustainable Infrastructure Networks, University of California, Berkeley.

(–24 feet mean sea level) on the island. Figure C maps the flooded areas to a height of –10 feet mean sea level. The green shaded areas on the maps are, for the most part, accessible within 10 minutes of travel time, whereas the red areas require closer to 30 minutes to reach. Accessibility to the south-central section of the island is quickly reduced, isolating those citizens from first responders within the critical first half-hour after an event occurs.

The findings of this type of GIS model go a long way to inform planners as they prepare to mitigate the effect of an event that might breach a levee, thus curtailing the road accessibility and response time that some areas may experience resulting from inundation. The results of this analysis can be dynamically updated as inundation conditions change and are added to the operational layers of the geodatabase for viewing within the COP of the situational awareness viewer.

Another analysis designed to identify the approximate extent of flood-water coverage was developed in Cobb County, Georgia, in response to a severe flood event in 2009. Managers in the EOC promptly requested a delineation of the flood boundary "to identify the approximate extent of floodwater coverage; identify and locate the affected parcels; determine the minimum, maximum, and mean water heights in the parcels; and identify and document completely and partially flooded properties."[30] Because Cobb County already had an enterprise GIS system in place as an integral part of the county's EOC, support GIS staff were able to quickly supply emergency managers with the information needed to ascertain the effects of flooding in the area. Maps and reports were generated on demand, which helped first responders, inspectors, and repair crews prioritize the areas to visit, and information was distributed to department managers and personnel via a web-based mapping application.

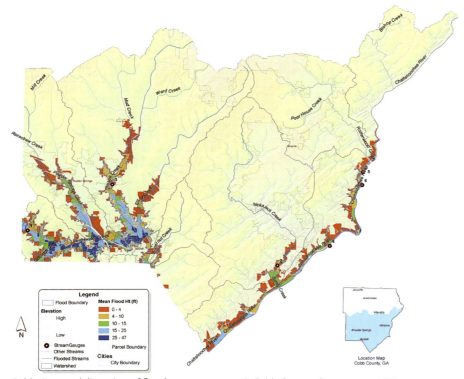

Cobb County delineation of floodwater coverage © Cobb County Government 2010.

Fire models and tools
Urban fire models

Fire models incorporate a wide range of variables into their analysis, depending on whether they are urban or rural in nature. In an urban setting, *standards of cover* goals set out the minimum deployment requirements necessary for a fire department to meet the public safety needs of a community. These goals are supported by response time and incident trend modeling. Response

30 Vinu Chandran, Seth Agyepong, and Tim Scharff, Cobb County GIS, "Getting Answers Quickly: Hydrologic Analysis and 3D GIS Improve Flood Response," *ArcUser Online*, April 2010, http://www.esri.com/news/arcuser/0410/cobbcounty.html.

time modeling analyzes the location of fire stations along a street network to determine the reach of fire response teams in all directions from all stations. The result is a map showing distance bands around each station that illustrates how long it would take in drive-time minutes for an emergency response team to reach each location in the event of an incident. Gaps in the bands determine those areas that are beyond the required reach of first responders, necessitating additional planning and mitigation measures to ensure access to those areas if an emergency should occur.

Accomack County Fire Minimum Time to Service

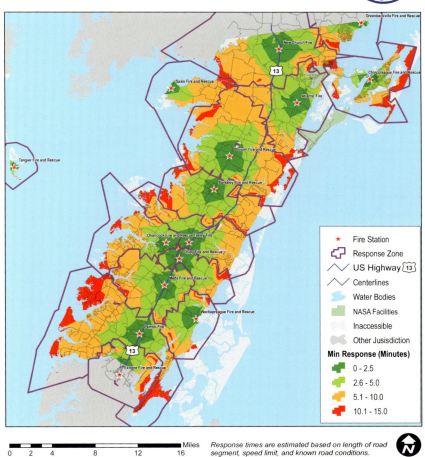

Minimum-time-to-service analysis Courtesy of Accomack County, Virginia. Department of Public Safety, Department of Planning, WorldView Solutions, Inc.

Incident trend analysis allows planners to map historical fire events in a community to see where they most frequently occurred in the past and how quickly first responders were able to arrive at the site.

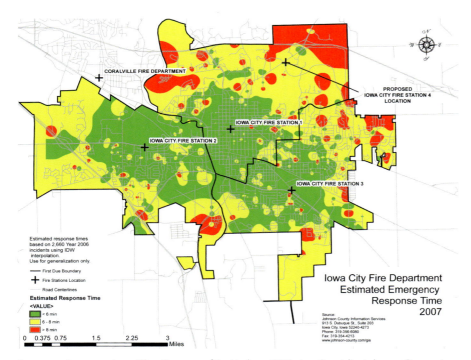

Response time map: Iowa City Courtesy of Jim Harken, GIS Project Specialist, Johnson County, Iowa.

In the case of the Baltimore City Fire Department, planners were able to determine from historical incident records exactly which responses failed to meet standards and where improvements were needed: "They could then begin to ask the all-important question of why some responses to calls failed to meet response time goals. A GIS analysis showed exactly which apparatus—a fire truck, fire engine, or chief—met or failed the standard for each incident and what the response times were for each vehicle."[31]

Filtering the historical calls database also enables fine-tuning of the analysis to show concentrations of only a certain type of event, such as arson, that occurred in a specific area of a community or at a particular time of day. The results of these GIS analyses can be effectively integrated into emergency

31 Jesse Theodore, "Baltimore City Fire Department Maximizes Manpower and Resources with Geospatial Technology," ArcWatch, April 2011, http://www.esri.com/news/arcwatch/0411/feature.html.

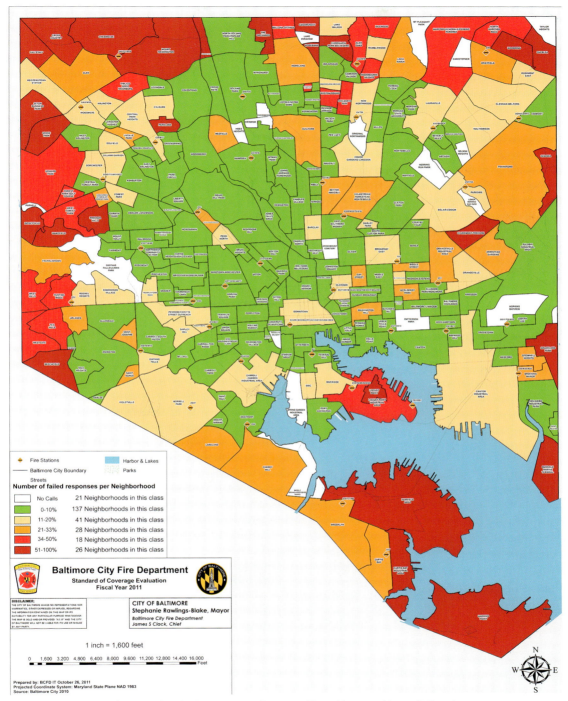

Response time map: Baltimore City Fire Department Courtesy of Peter Hanna and James R. Potteiger.

planning and mitigation plans for a community. Studying the spatial distribution of resources and events enables the most efficient and effective allocation of firefighting personnel and apparatus to ensure the greatest public safety.

Wildland fire models

Within a rural or undeveloped region, wildland fire models simulate fire behavior and growth based on the topography, fuels, and weather of an event. For planning purposes, the most commonly used models are the set of spatial fire behavior systems supported by the US Forest Service that integrate with a GIS to provide maps and tables depicting the perimeter, travel time, intensity, and probability of a fire spreading across a landscape.

- Rothermel's 1972 method for modeling the spread of wildfire is still the most widely used approach to fire planning and management. This model, coupled with Huygens' principle of describing the movement of light waves, uses points to define the fire front as independent sources of small elliptical wavelets.[32] The shape of the fire front can be manipulated to account for variations in slope, wind, and other environmental considerations.

- FARSITE (Fire Area Simulator) is a Windows-based fire-simulation application that interfaces with ArcGIS to model how fire will spread over time and space under certain topographic and weather conditions given particular fuel inputs. It uses Rothermel's method to produce tabular and map data that is used primarily for long-term fire prediction analysis and is effectively integrated into the planning and mitigation phases of emergency management to assess fire suppression methods and preparedness.

- Wildfire Analyst is a fire-simulation tool that interfaces with ArcGIS to model fire behavior within an operational context. It provides real-time analysis of wildfire spread, fire behavior, suppression capacity, evacuation time, and impact analysis during an incident. Simulations run in less than 2 minutes, providing real-time capabilities to adjust simulations with observed data and proposed suppression activities. The tool is fully integrated into the ArcGIS common operating platform, making it available to the firefighter in the field, as well as the fire behavior analyst expert at the command center. This integration into the operations phase of emergency management greatly enhances the situational awareness of a fire event as it unfolds, providing commanders and first responders with a common view of relevant real-time information.[33]

---

32 "Modeling Fire Growth (technical documentation)," FireModels.org, Fire Behavior and Fire Danger Software, Missoula Fire Sciences Laboratory, http://www.firemodels.org/downloads/ farsite/webhelp/technicalreferences/tech_modeling_fire_growth.htm.

33 "What Is Wildfire Analyst?" DTS/Tecnosylva, http://www.wildfireanalyst.com/index.php/ wildfire-analyst/what-is-firesponse.

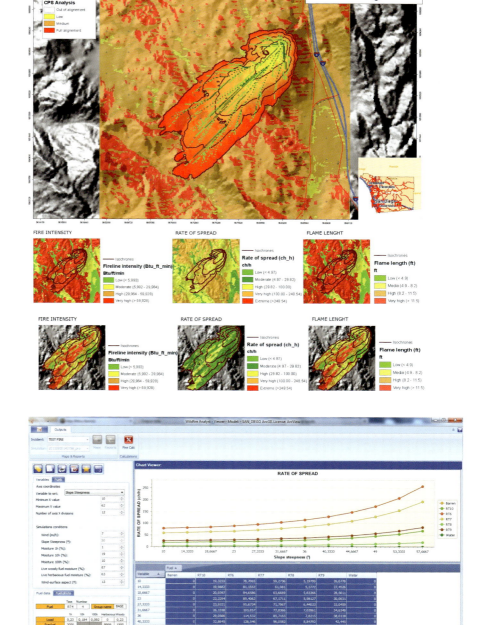

Real-time fire-spread modeling: Maps, tables, and charts Courtesy of Texas Forest Service.

Summary

The advantage that GIS brings to emergency management rests not only in its ability to show all response personnel simultaneously *where* an event is occurring, but also *how* it is unfolding at the time or expected to unfold in the future. Vulnerability assessments of a community identify those areas most likely to suffer the greatest loss should a catastrophic event occur. GIS-based models that simulate the expected consequences of particular hazardous events show the extent to which people and places will be affected should a particular event occur.

Although these analyses have traditionally remained within the domain of scientists and highly skilled GIS analysts, their findings can now be effectively included in the enterprise GIS of a local government and emergency management agency. Integrated as operational layers, this information can be on hand as emergency managers plan for and mitigate the expected effects of such catastrophic events as earthquakes, floods, hurricanes, fires, hazardous spills, and other technological disasters. This mission-critical information can now also be readily accessed by emergency managers as they gauge the real-time effects of an event as it unfolds by using web-enabled operational tools embedded in COP applications for emergency response and planning.

Chapter 4: Field operations

Incidents don't occur in the EOC, they happen in the field. The ability to communicate in real time with first responders and recovery personnel as they walk the landscape of a catastrophic incident greatly enhances the overall situational awareness of an event, saving lives and securing property more quickly and most effectively. An EOC equipped with a GIS-based data management system provides an integrated common operating platform that enables emergency personnel to direct effective response and recovery operations. Comprehensive situational awareness requires real-time, or near real-time, information integration as an incident evolves. Gathering data and information at the event site is the only way to truly gauge the extent of impact and determine what resources are needed and where they are best deployed. The use of GIS-ready mobile devices during incident response, and later during recovery operations, is the final operational component of complete situational awareness of an event.

In a fully integrated common operating platform, information shared between event partners flows both into and out of the central enterprise geodatabase. Data sent *from* the geodatabase out to the field includes the GIS layers that provide geographic reference information to on-site field personnel. The location of fire hydrants, the layout of building footprints, the closest medical facilities, and places where hazardous materials may be stored are all important pieces of information first responders need to support decision making. Recovery personnel also require information such as transportation networks, utility lines, parcel footprints, imagery, and property ownership to effectively assess the extent of damage occurring from an event.

Conversely, data sent *to* the geodatabase from the field includes real-time updates of current on-site incident conditions. This includes both rapid and detailed damage assessments, current operational locations, search-and-rescue missions, and recovery requirements after the incident. This mutual data-sharing workflow among the EOC, incident command, and field operations personnel provides the complete information interoperability necessary to enable comprehensive situational awareness in the event of a catastrophic event.

Field data collection, however, has typically been marred by communication incompatibilities and inadequate assessment tools, delaying the exchange of information. The use of more versatile and user-friendly mobile GIS devices and better-designed operating systems now provide field operations personnel access to the same real-time information on handheld units that is available in the EOC. Custom data entry applications that facilitate real-time updating and reporting of critical information are now readily available via a host of mobile GIS platforms and devices that serve a range of operational needs throughout an emergency management and operations enterprise.

Mobile GIS

Mobile GIS extends the reach of GIS from the EOC out into the field and back again. GIS-enabled handheld devices now allow response and recovery personnel to view, collect, and update geographic information as an event unfolds. That same information gathered in the field can simultaneously be viewed by duty officers in the EOC and operations personnel in the field and at incident command posts as they make decisions that save lives and secure property.

The mobile GIS environment consists of a few key components. First and foremost is a mobile device equipped with a satellite-based Global Positioning System (GPS) receiver. This capability aligns the data collected with its geographic location. GPS measures the distance of locations on the ground to a constellation of satellites orbiting the earth. By calculating the time it takes for a signal to be transmitted between the receiver and the satellites, the system can compute the coordinate location of the receiver situated on the ground. The mobile GIS device then stores this location and is ready to accept additional information the user enters about that location.

In order to enter descriptive information about a particular location, the device must be equipped with a software application that integrates the GPS capability into a GIS environment. This mobile application is authored at the desktop and embeds directly into the GIS workflow of an organization. It contains the map windows and data entry forms where field personnel can enter and edit information that describes a site.

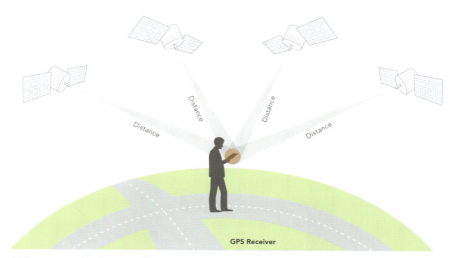

GPS measures the distance of locations from a constellation of orbiting satellites Esri.

Esri offers a range of GIS software applications for mobile devices, including ArcGIS for Windows Mobile, ArcPad, and ArcGIS for Smartphones and Tablets:

- ArcGIS for Windows Mobile complements both ArcGIS for Desktop and ArcGIS for Server with task-centric field applications for vehicle-mounted Windows touch devices and handheld Windows Mobile devices. It is designed for non-GIS professionals in medium to large organizations that perform simple data collection and field inspection projects.

- ArcPad is a map-centric, Windows-based mapping and field data collection application designed for GIS professionals in small- to medium-sized work groups who typically perform ad-hoc data collection. It includes advanced GIS/GPS editing functionality to facilitate post processing and also supports related tables. ArcPad projects can be customized using scripts and applets available in ArcPad Studio.

- ArcGIS for Smartphones and Tablets extends the use of GIS to a much wider audience. This mobile application puts GIS into the hands of anyone with a Windows Phone 7, Android, or Apple iOS device. Users can easily navigate maps, collect and report VGI, and perform GIS analysis. An organization can access and share its enterprise GIS via ArcGIS Online or ArcGIS for Server and build custom mapping applications that meet the

The ArcPad interface Courtesy of Trimble.

specific needs of its users. Embedded GPS capabilities of most
smartphones, coupled with the mobile ArcGIS application, turn
each handheld unit into a data device that can capture, store,
and upload photos and videos taken at the site, as well as geocode
information entered into an online damage assessment form.

A GPS-ready device equipped with a mobile GIS software application
can access an extensive set of geographic information to use as reference data for its mapping applications. The enterprise local government
geodatabase designed in the data management phase of GIS development
can be leveraged by a mobile GIS application through data synchronization between the client and server devices via wireless communication.
This same data synchronization serves data collected in the field back
to the central enterprise geodatabase to enable a completely connected
common operating GIS platform. This interoperability between data and
devices, and the EOC and field personnel, greatly enhances the situational
awareness of all event partners managing a catastrophic incident.

Author, Serve, and Use

Author content using ArcGIS for Desktop

Serve authored content using ArcGIS for Server

Web Services

Desktop

Windows Smartphone

Mobile Applications

Tablet PC

Pocket PC

Mobile ArcGIS system configuration Esri.

Mobile GIS in emergency field operations: Response

Mobile GIS plays a pivotal role in the management of emergency operations both during and after an event occurs. During an event, incident commanders and first responders can use mobile GIS to guide their search-and-replace operations in a changed landscape. Accessing basemap layers that reference pre-event conditions can enable first responders quick and safe entry into an area even after preexisting access routes become inoperative.

As responders navigate the affected landscape, they can report real-time conditions as operational layers back to the enterprise geodatabase so incident commanders and other field personnel deployed in the area are equally informed. This data flow ensures the timely and accurate situational awareness for all operational personnel during the life cycle of an event.

The tactical value to response personnel of a real-time interactive mobile GIS data application cannot be overstated. Having relevant and up-to-date information in hand as an event unfolds provides operational personnel timely and accurate information to support critical decisions. Timely decision making based on accurate information saves lives and reduces damage to property and natural resources. Many fire departments across the United States are deploying these types of mobile GIS systems. The Turlock Fire Department in California is in the process of outfitting its five engines with technology designed to allow firefighters to more quickly and efficiently respond to emergencies. The enhanced mapping system will enable firefighters to more efficiently find their route to affected locations and locate fire hydrant positions relative to these sites. Before the new system was installed, dispatched firefighters had to search through binders of hard-copy documents to gather relevant information.[1]

The City of Marietta Fire Department in Georgia, in partnership with Geographic Information Services, Inc. (GISi), has also successfully deployed a state-of-the-art mobile mapping application that routes and reroutes fire trucks on the fly.[2] This ArcGIS for Windows Mobile application is a powerful and highly customized environment that communicates in real time with the city's enterprise databases, eliminating the need for manual data updates. It includes automatic vehicle location (AVL), routing and rerouting capabilities, and access to building plans in a single, easy-to-use application that displays this integrated data in an online map window.

Routing and rerouting capabilities are critical in a response operation as previously accessible routes and access points become blocked by changing conditions. It is also useful in mutual aid situations, when firefighters are called to assist in a neighboring jurisdiction where they are unfamiliar with the streets and neighborhoods.

Another important feature of this system is its ability to simultaneously show multiple views of an area as responders approach a site. Multiple-view displays include the locations of fire hydrants as well as a 360-degree view of individual buildings and the areas around them,

---

1 Steve Milne, Capital Public Radio, Sacramento, California, September 22, 2011.
2 Matteo Luccio, "Mapping Application Makes Accurate Data Available in Real Time—Mobile GIS Helps Firefighters Respond More Effectively," *ArcWatch*, June 2010, http://www.esri.com/news/arcwatch/0610/marietta-fire-dept.html.

City of Marietta mobile mapping application Courtesy of City of Marietta, Georgia.

including floor plans and plat images. Another feature alerts users to the presence of hazardous materials or chemicals in a building, yielding better tactical decisions throughout the response phase of an incident.

The use of mobile GIS in a wildfire scenario is equally essential. Before mobile GIS was integrated into the real-time tactical operations of wildfire response and recovery, firefighting strategies and operations originated from a command post outside the affected area and were periodically updated with information gleaned from field sources, aerial reconnaissance, and satellite imagery. In 2003, when communications from these field sources could take up to 12 hours to update, firefighters on the front line of the Southern California wildfires found that conditions on the ground differed greatly from the latest updates they received from the command post. A mobile GIS trial was then deployed, pairing mobile GIS devices with GPS-ready helicopters that flew the perimeter of an active fire and transmitted their coordinates back to the enterprise GIS, where supervisors and command officers were then able to see the current extent of the raging firestorm.

This 2003 mobile GIS trial laid the cornerstone for the expansive development and deployment of real-time fire mapping that is widely used today.

By 2007, a fully integrated common operating platform was assembled in Southern California in response to another outbreak of fires across the same Los Angeles, Orange, Riverside, San Bernardino, San Diego, and Ventura counties. This GIS-ready platform was able to support mobile operations that transmitted real-time fire perimeter coordinates back to the command center where interagency teams were able to make timely decisions that saved lives and secured property across the region. These updated fire maps were also available online for the first time via an ArcGIS Explorer application that displayed the current perimeter and conditions of the wildfire.[3] It also provided a link to a detailed fact sheet for each fire incident available on the InciWeb website, an interagency all-risk incident information management system with continuous real-time fire updates across the United States.[4]

To supplement the TxWRAP web portal, Texas Forest Service has partnered with DTS (Orlando, Florida) to extend capabilities with a mobile application that operates on any smartphone or tablet platform. TxWRAP will automatically determine if the user is operating a standard web browser or a mobile device and serve the appropriate application interface to the user. The mobile version enables users to interact with key risk-assessment data layers, including fire behavior analysis and landscape characteristics.

In addition, active weather conditions and predictive services data is available in concert with fire incident locations, updated every 15 minutes, and MODIS fire-detection data. This integration of operational fire incident data with risk conditions provides the information necessary for field staff to obtain situational awareness quickly while responding to incidents.

To aid staff in using the data, a suite of distance and area tools are embedded in the application to enable measurements from the map (or current GPS location) to gauge the extent of spread and perimeter of active fires. Landscape characteristics maps are also included that identify surface fuels, vegetation, rates of spread, flame length, canopy fire potential, housing density, and the values-at-risk. Additional tools offer predictive services that identify the wind speed and direction, relative humidity, precipitation, drought index, and forecasted and observed fire danger. TxWRAP mobile helps inform first responders of current conditions using the agency's investment in a centralized risk-assessment portal.

Tools such as these put the power of real-time mobile GIS into the hands of incident commanders, first responders, field personnel, and ordinary citizens as they work together to restore safety and security to threatened communities.

---

3 Monica Pratt, "Déjà Vu: Four Years Later, GIS Use in Fighting Fires Greatly Expanded," *ArcUser Online*, October–December 2007, http://www.esri.com/news/arcuser/1207/dejavu.html.
4 "InciWeb: Incident Information System—Current Incidents," US Forest Service, http://www.inciweb.org/.

Texas Wildfire Risk Assessment Portal displayed on a smartphone

Courtesy of Texas Forest Service.

Mobile GIS in emergency field operations: Recovery

One of the greatest and most immediate advantages of mobile GIS is its ability to quickly automate and integrate the damage assessment information a community needs to recover from a catastrophic event. Recovery personnel can effectively use mobile GIS to perform accurate inspections as they scan an affected area to determine the extent of damage. The data, including form-based information fields and geotagged photos, can be seamlessly uploaded in real time via wireless communications to the common operating GIS platform at the EOC for immediate use by incident commanders and duty officers.

In the following example, ground inspectors are in the process of surveying buildings in an area affected by wind damage from a recent hurricane.

The map at the EOC shows the buildings that have already been surveyed, with their footprints symbolized in red (severe damage), orange (moderate damage), yellow (light damage), green (no damage), or gray (not surveyed).

Damage assessment inspection survey using an integrated COP and mobile GPS device

Map by Esri; data courtesy of LOJIC, Louisville, Kentucky; device courtesy of Trimble.

The inspectors are equipped with handheld mobile GIS devices that, whether connected to the network or not, are able to see the exact same map that the personnel at the EOC use to manage the recovery operation. Once the gray building is selected for inspection, the damage assessment form is loaded onto the mobile screen and data is entered identifying the building and material type and the condition of the structure. This is the same type of information traditionally collected on a paper form, but the handheld GIS device eliminates the need to transcribe that information into a database once back at the incident coordination center. This not only expedites the information workflow, but also reduces the risk of data entry errors.

Once surveyed, the information is updated on the mobile device, and then synchronized and posted to the enterprise geodatabase at the EOC.

The update is now shared via the situational awareness map viewer with all event partners who are integrated into the common operating GIS platform.

This damage assessment technology is available across the full suite of mobile GIS devices and applications. The example shown is an ArcGIS for Windows Mobile application deployed in the field on a handheld mobile GPS device. This same integrated application is also supported by the web-based

Damage assessment inspection data entry form on a mobile GPS device Device courtesy of Trimble.

Damage assessment inspection survey synchronized back to the COP from a mobile GPS device
Map by Esri; data courtesy of LOJIC, Louisville, Kentucky; device courtesy of Trimble.

situational awareness map viewer back at the EOC. Damage assessment applications also exist on common mobile devices that are equally integrated into the GIS common operating platform. Many agencies use ArcPad to develop custom damage assessment applets that can be mounted on tablets and smartphones, enabling recovery teams deployed in the field to access damage assessment forms registered to their agency's enterprise database.

This technology became a lifeline for recovery personnel facing the challenge posed by the 2009 Australian bush fires that devastated more than 220,000 hectares of land in the state of Victoria: "Entire towns and communities ceased to exist. Familiar landmarks, such as street signs, mailboxes, residences, and businesses, were reduced to smoldering rubble. The number of missing or unaccounted-for individuals continued to increase, along with the growing numbers of deaths during this unprecedented catastrophe."[5] At the start of the recovery operation, traditional search-and-rescue teams were initially deployed, with each affected property fully inspected out to a 50-meter extent, including all outbuildings, drain pipes, culverts, and mine shafts—anywhere someone might seek protection from the approaching firestorm. Each inspection was logged onto color-coded paper forms, which were submitted at the end of the day to a central pool at the Rescue Coordination Center 60 kilometers away. The information was entered manually into a database that updated tactical maps showing which areas had been adequately searched. These maps were then used to inform command and field personnel on a regular briefing rotation that sometimes took up to 48 hours to update.

Recognizing that the lag time of this workflow could seriously jeopardize the success of the recovery operation, a collaborative group of GIS analysts and programmers swiftly developed an ArcPad custom applet for rapid damage assessment. By transferring the same information from the paper forms to the handheld mobile GIS unit, recovery personnel familiar with the traditional paper-based data entry environment were able to use this digital applet with little or no training or GIS expertise. The rapid recovery team was then able to search almost 5,800 properties and successfully transmit the records via a 3G cellular network to the server at the coordination center in a fraction of the time it would have taken had they not been using the ArcPad application.[6] Mobile GIS played a pivotal role in the success of this mission and the overall situational awareness of all partners in this emergency operation.

5 "Mobile GIS Aids Victoria Bushfires Search Operations," *ArcNews Online*, Summer 2009, http://www.esri.com/news/arcnews/summer09articles/mobile-gis-aids.html.
6 Ibid.

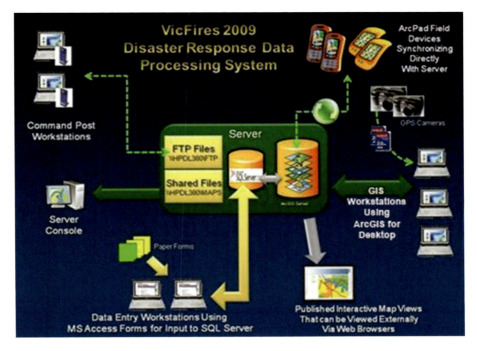

Victoria bushfires disaster response data processing system using ArcPad mobile GIS in an integrated common operating platform Esri.

More recently, similar mobile GIS technology was successfully deployed after the Tuscaloosa, Alabama, tornado killed forty-one people and left behind a wide swath of devastation and destruction in the spring of 2011. During the initial structural damage assessment response to this catastrophe, recovery personnel used the Geocove mobile GIS system, ARM360[7] to "make sense of the ruins left behind by the storms."[8] Maps and reports of damaged structures and loss estimates were automatically updated as assessment teams surveyed the damage from the field, offering the incident management team better operational awareness (see page 114).

Mobile GIS was also used in the damage assessment and debris removal operation following the devastating tornado that struck Joplin, Missouri, in 2011. Like other natural disasters, much of the pre-event transportation and communications infrastructure along the path of the tornado suffered almost total destruction. Maps of pre-event roadways and access points could not be used to route recovery personnel into affected areas. Vehicles

7 "ARM360," Geocove, http://www.geocove.com/arm360.
8 Gail Short, "GIS Proves Beneficial for Disaster Recovery—Digital Maps Help Responders Find Their Way in the Ruins," *American City and County*, July 1, 2011, http://americancityandcounty.com/technology/gis_gps/gis-disaster-recovery-201107/.

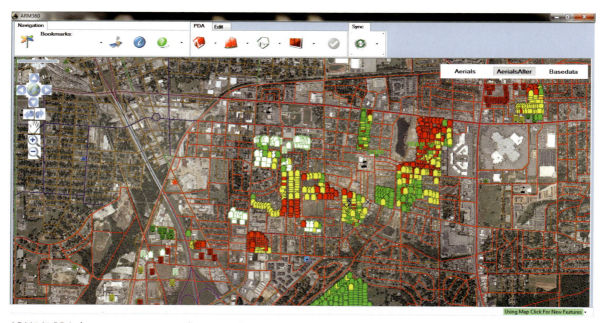

ARM360 PDA damage assessment application and in-progress field awareness map of Tuscaloosa
tornado damage   Courtesy of Karyn Tareen, Geocove.

equipped with GPS-enabled devices allowed "personnel to find their exact location in the field when all existing land markers had been destroyed. They could send their location along with an initial triage report immediately back to the EOC or incident command post. With numerous devices deployed, the information was wirelessly transmitted and displayed on the primary maps within the common operating picture."[9] Once the precise locations of recovery vehicles were established, a web-enabled mobile GIS application was deployed to "estimate debris type and amount. The application was used to identify where certain types of debris could be routed to and piled (temporarily), where debris could be piled and burned, and where environmentally harmful debris could be treated, managed, or removed."[10]

To date, the largest deployment of mobile GIS in the field was in support of the effort to protect wildlife and natural resources after the Gulf oil spill. Planning staff members, working in EOCs in five different states, were kept informed of oil spill conditions and the effects on wildlife by transmitting data collected in cellular dead spots of the Louisiana marshes through an Inmarsat Broadband Global Area Network (BGAN) terminal.[11] The BGAN device is a Wi-Fi-enabled, compact, portable data/voice satellite transceiver with global Internet coverage. Rugged PDAs running ArcGIS Mobile mapping applications synchronized with an ArcGIS Server instance in Texas directly from affected locations throughout the Gulf. This deployment of mobile GIS technology clearly demonstrated the capabilities of ArcGIS used as a complete system for data management, planning and analysis, field operations, and situational awareness (see page 116).

Damage assessment data is vital to starting long- and short-term recovery. It also supports the requirements for mandatory reporting to government agencies for disaster relief funds that begin to restore normalcy and security to affected communities. Traditional reporting of damage to affected properties takes extended periods of time and exhaustive dedication of resources to complete. As highlighted in the case of the Australian bush fires, the transition from paper-based conventional damage assessment to a mobile GIS system significantly expedited the collection of data, which enabled recovery personnel to locate missing persons more efficiently. The added benefit of already having the damage assessment data gathered in the field pooled in the enterprise GIS system is that it enables the quick and efficient compilation of this information for reporting. It used to take

9 Eric Holdeman, "Russ Johnson Talks Esri and All Things GIS," *Emergency Management*, November 7, 2011, http://www.emergencymgmt.com/disaster/Russ-Johnson-Talks-Esri-GIS.html.
10 Ibid.
11 "Volunteered Geographic Information Plays Critical Role in Crises-Redlands GIS Week," *ArcWatch*, March 2011, http://www.esri.com/news/arcwatch/0311/power-of-vgi.html.

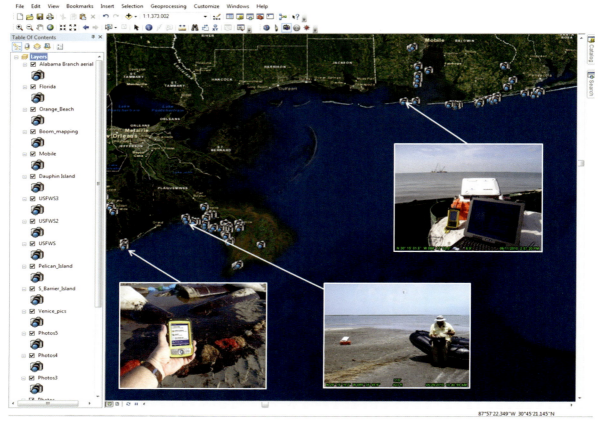

Mobile GIS used to protect wildlife and natural resources after the Gulf oil spill Photographs and graphic compilation by Tom Patterson for Esri; data from i-cubed, information integration and imaging, LLC—distributed through i-cubed's DataDoors Archive Management, http://www.datadoors.net.

a dedicated pool of personnel to compile the data for these critical reports. Now that the data is in an integrated common operating platform, it can be compiled and reported as it is collected. Expediting this reporting function means that funding support from government agencies gets into the hands of the people who need it most, faster and more efficiently than ever.

Confirmed Damaged Structures

3/11/2009

If your property has suffered damage due to the chemical release and does not appear on this list or the map, please call the Emergency Management office at 502-555-1212. Please do not reenter any structure until it has been deemed safe by the health department.

APN	Street Address	City	Zip	Type	Damage
014B00820000	1311 W MARKET ST	LOUISVILLE	40203	Residential	Moderate
014G00250000	1351 W MUHAMMAD ALI BLVD	LOUISVILLE	40203	Business	Light
014A00500000	1612 W MARKET ST	LOUISVILLE	40203	Residential	Light
014A01290000	1637 W MARKET ST	LOUISVILLE	40203	Residential	Moderate
014B01180000	1214 W MAIN ST	LOUISVILLE	40203	Residential	Heavy
014C02120000	1030 W MARKET ST	LOUISVILLE	40203	Residential	Heavy
014G01750000	1420 W JEFFERSON ST	LOUISVILLE	40203	Business	Moderate
015F01200000	1618 COLUMBIA ST	LOUISVILLE	40203	Residential	Moderate
015F00630000	1723 ROWAN ST	LOUISVILLE	40203	Residential	Light
015F00750000	1709 ROWAN ST	LOUISVILLE	40203	Business	Moderate
014F01420000	1518 CEDAR ST	LOUISVILLE	40203	Business	Heavy
014B00250000	1328 W MARKET ST	LOUISVILLE	40203	Residential	Light
015A00610000	1020 W MARKET ST	LOUISVILLE	40203	Residential	Light
015A00470000	1601 W MAIN ST	LOUISVILLE	40203	Business	Light
014A01260000	1704 W MAIN ST	LOUISVILLE	40203	Business	Moderate
014A01620000	1537 W MARKET ST	LOUISVILLE	40203	Business	Light
014B01010000	1205 W MARKET ST	LOUISVILLE	40203	Business	Light
014C01090000	1145 W MARKET ST	LOUISVILLE	40203	Residential	Moderate
014A00660000	1516 CONGRESS ST	LOUISVILLE	40203	Business	Moderate
014G00930000	1328 W LIBERTY ST	LOUISVILLE	40203	Residential	Heavy
014A01300000	1635 W MARKET ST	LOUISVILLE	40203	Residential	Light
015A00640000	1528 ROWAN ST	LOUISVILLE	40203	Business	Moderate
015F01260000	212 N 16TH ST	LOUISVILLE	40203	Business	Light
015A00810000	1721 CROP ST	LOUISVILLE	40203	Residential	Light
014A01270000	1700 W MAIN ST	LOUISVILLE	40203	Business	Moderate
014A01480000	113 S 17TH ST	LOUISVILLE	40203	Residential	Moderate
014A00820000	1505 CONGRESS ST	LOUISVILLE	40203	Residential	Heavy
014G00920000	1330 W LIBERTY ST	LOUISVILLE	40203	Business	Light
015F01110000	1607 ROWAN ST	LOUISVILLE	40203	Business	Moderate
015A00830000	1520 BANK ST	LOUISVILLE	40203	Residential	Light
014A00650000	1511 W JEFFERSON ST	LOUISVILLE	40203	Residential	Light
014F00560000	1626 CEDAR ST	LOUISVILLE	40203	Residential	Moderate
015B00740000	223 N 15TH ST	LOUISVILLE	40203	Residential	Moderate
015A00300000	1710 ROWAN ST	LOUISVILLE	40203	Residential	Heavy
015B00080000	1301 W MAIN ST	LOUISVILLE	40203	Residential	Light
014A01160000	114 S 17TH ST	LOUISVILLE	40203	Residential	Heavy
014A00810000	1509 CONGRESS ST	LOUISVILLE	40203	Business	Light
015F01250000	1608 COLUMBIA ST	LOUISVILLE	40203	Residential	Light

Detailed damage report ready for submission to government agencies for disaster relief funds

Data courtesy of LOJIC, Louisville, Kentucky.

Summary

The connection between field personnel on site and emergency operations managers at the command center is a critical relationship in the emergency management workflow. To be able to support decisions at the EOC that have great impact on how a response and recovery operation unfolds, emergency management officials depend first and foremost on information about what is occurring on the ground in real time. A fully integrated common operating GIS platform enables this vital link in a real-time geographic context that supports critical decision making that saves lives, property, and natural resources.

The full suite of mobile GIS devices and applications supports emergency management personnel during both the initial response phase and later recovery operations of an event. ArcGIS for Windows Mobile, ArcPad, and ArcGIS for Smartphones and Tablets integrate seamlessly with the emergency management workflow to enable a complete operational environment for managers and field personnel to coordinate lifesaving activities. Traditional response and recovery methods are adapted to these digital environments to help non-GIS personnel easily transition into the workflow. The confidence that this technology gives to emergency operations personnel greatly contributes to their effectiveness and success. This complete awareness of all factors affecting a community before, during, and after an event enables good planning and quick response and recovery that saves lives and secures property.

Chapter 5: Situational awareness

The previous chapters of this book discussed the depth and breadth of the critical components of the ArcGIS common operating platform that enable comprehensive situational awareness within the scope of emergency management. The goal of all data management, planning, analysis, and field operations activities is to provide the greatest understanding of the current situation, particularly of real-time events as they unfold at an affected site. The key to achieving this goal, second only to cooperation among event partners, is a clear and well-designed workflow supported by a fully integrated and robust system architecture that effectively delivers information to decision makers, translating it from raw data into actionable intelligence.

As highlighted many times in earlier discussions, a web-enabled COP supported by a geodatabase of both base and operational data layers, combined with well-designed analytical tools, facilitates effective decision making with timely and accurate information. A fully configured GIS platform that supports shared mission-based situational awareness enables emergency management personnel to act quickly, coordinate more effectively, and target actions on the most critical issues. Many public safety organizations have implemented GIS-based situational awareness systems that have resulted in greater awareness, more effective response, and better incident management coordination. An example of this type of GIS-based COP viewer workflow includes the application developed by the Fire Department of New York (FDNY).

Not only is this system used for response, but it is also used early in the planning process when major public events are scheduled in the city. The system is then used to monitor, manage, and adjust resource allocation needs as the event unfolds.

FDNY COP Critical Response Information Management System (CRIMS) Courtesy of Fire
Department City of New York–Geographic Information System (GIS) Unit.

The system is configured to integrate and display calls from the CAD system
as they are received. Calls tagged as significant events are then entered into
the COP address locator. Once the location is identified, supporting web services linked to the map are called upon to provide a complete set of documents
that define the event. These include street-level and oblique photography of
the site and surrounding area and a building profile describing any known previous incidents or on-site hazardous conditions. Finally, all of this information
is distributed via email within the FDNY intranet to critical staff members
within minutes of recording the incident. The awareness that this GIS-based
system offers provides fire personnel and duty officers immediate and relevant
information that directly supports their work to protect the community.

This first-generation situational awareness delivered through a single
COP has changed the way many agencies respond to emergency incidents. It has, however, evolved to a more comprehensive, better-designed
mission-specific pathway through the life cycle of an event. The great
extent of data, coming from numerous sources within an agency and
supplemented by real-time data from the field, can overwhelm critical
decision makers unless that information is well presented and clearly
understood. The single map COP, although comprehensive by definition,
may be "too much information" for a particular event partner with a
mission-specific task at hand. The ability to carve out relevant workflows
and apply filters that deliver complex data and tools based on a user's
role goes much further to enable effective situational awareness than
a throng of seemingly disparate and confusing data layers and tools on
a common map. The ArcGIS common operating platform now provides

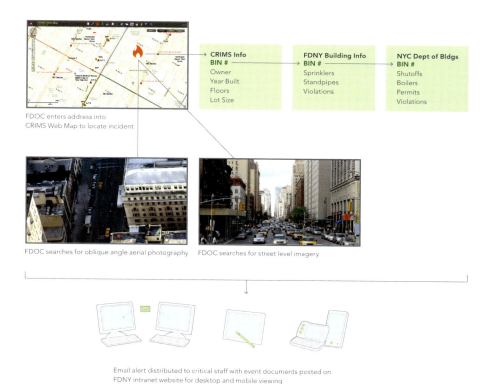

Incident Happens
(Fire, Suspicious Package, etc.)

CRIMS Info
BIN #
Owner
Year Built
Floors
Lot Size

FDNY Building Info
BIN #
Sprinklers
Standpipes
Violations

NYC Dept of Bldgs
BIN #
Shutoffs
Boilers
Permits
Violations

FDOC enters address into
CRIMS Web Map to locate incident

FDOC searches for oblique angle aerial photography

FDOC searches for street level imagery

Email alert distributed to critical staff with event documents posted on
FDNY intranet website for desktop and mobile viewing

FDNY situational awareness information workflow Courtesy of Fire Department City of
New York–Geographic Information System (GIS) Unit; photographs © Thorsten Nieder/Fotolia
and © Marcel Schauer/Fotolio.

not only the necessary data and analysis foundation from which to
serve and consume actionable intelligence, but it allows custom pathways and mission-specific views of data and access to relevant tools
and applications that better serve those making critical decisions.

Hence, the focus of the GIS solution for situational awareness is evolv-
ing from the single common operating *picture* to the integrated common
operating *platform*. The universal requirement remains access to data,
services, and tools. All personnel within a mission space have access to
the same base data, as discussed in chapter 2. The operational layers,
tools, and services, however, are best deployed when they are prescrip-
tive and based on mission needs. Custom-designed mission-specific

operating pictures, or intelligent digital maps, enable each unit within a public safety organization to focus its attention and response efforts on those tasks and decisions that define its scope of work. This task-centric approach to the COP capitalizes on ease of use and operational readiness to make it the preferred workflow for successful emergency response.

ArcGIS for Emergency Management[1]

The ArcGIS tools and applications are designed within the emergency management context to deliver content and allow information flow across the organization in a targeted and meaningful manner. In place of one COP, multiple mission-specific viewers, aligned with the NIMS, can deliver data and tools based on the ICS framework. Baseline configurations for command, operations, planning, logistics, public information, and for each of the National Response Framework (NRF) ESFs provide a starting point for better situational awareness.

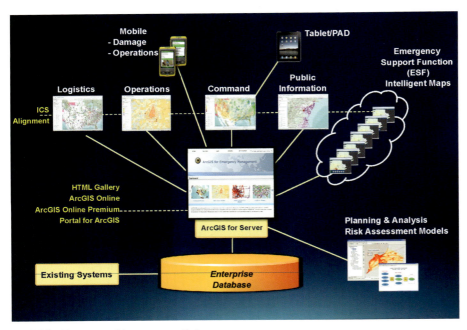

ArcGIS for Emergency Management Esri.

---

1 Adapted from Esri Public Safety white paper, "ArcGIS for Emergency Management," included in appendix A.

Central command dashboard

At the highest level, the system is "driven" by a *command viewer* that provides an executive-level (overall) view of the jurisdiction and current operations. This command viewer is dynamic, providing the most current and comprehensive status of events for life-cycle emergency management and general department head staff as they monitor the community's situation. It includes major hazard feeds (e.g., weather, earthquake, tsunami, and hurricane feeds), as well as connections to real-time crisis and CAD systems for updates on law enforcement, medical, fire, and hazmat incidents. This high-level viewer is not overloaded with the full suite of data and analytical tools available, but rather is configured as an easy-to-use picture of existing conditions that provides basic reporting and markup capabilities.

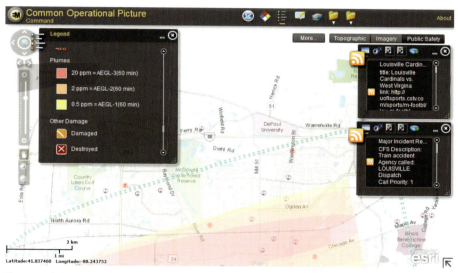

Command viewer Map by Esri; data courtesy of City of Naperville; RSS icon courtesy of Mozilla Foundation.

The command viewer is accompanied by other focused map viewers that provide critical information for other roles in the ICS framework, such as operations, planning, logistics, and public information.

Operations/tactical planning view

The operations/tactical planning view displays response and tactical operations activities that a jurisdiction oversees. It is intended for operations staff members that monitor and support response and recovery operations. It contains comprehensive base and operational data layers, and integrates

GIS analytical tools to enable comprehensive support of tactical operations. Its primary focus is on the NRF ESF 5—Emergency Management function, which includes coordination of incident management and response efforts, issuance of mission assignments, deployment of resources and human capital, incident action planning, and financial management. It is supported by an additional set of nested mission-specific map viewers that align to the remaining ESFs, to include the following:[2]

- ESF 1—Transportation
- ESF 2—Communication
- ESF 3—Public Works
- ESF 4—Firefighting
- ESF 6—Mass Care
- ESF 7—Logistics/Resources
- ESF 8—Public Health
- ESF 9—Search and Rescue
- ESF 10—Oil and Hazardous Materials
- ESF 11—Agriculture and Natural Resources
- ESF 12—Energy
- ESF 13—Public Safety
- ESF 14—Long-Term Recovery
- ESF 15—External Affairs

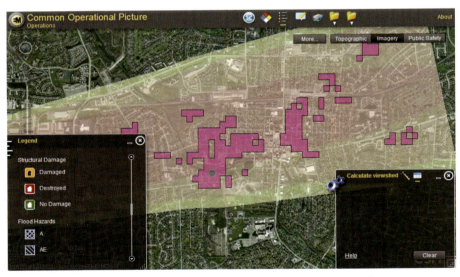

Operations/tactical planning view Map by Esri; data courtesy of City of Naperville.

2 A complete list and description of all ESFs and accompanying data and tools can be found in the Esri Public Safety white paper, "ArcGIS for Emergency Management," included in appendix A.

The tools most important at this level of operations are those that support incident-based editing and red lining, allocation of shelter assignments, damage assessment and debris removal monitoring, search-and-rescue grid and crew tracking, evacuation logistics, and updates on critical infrastructure and known resources in the region. Cumulatively, these viewers help decision makers determine priority recovery actions across the jurisdiction, as well as maintain continuity of government and community operations.

Logistics/resource view

The logistics/resource view is a mission-specific map view that focuses exclusively on the availability and management of response and recovery resources. Its primary goal is in support of local and regional incident management and recovery, as well as mutual aid support from neighboring jurisdictions. The availability, tasking, and routing of resources are managed through interaction with incident data, including the location of incident posts, staging areas, shelters, hospitals, and search-and-rescue personnel and equipment. The map is supported with routing capabilities that integrate live transportation updates on road access and pathway options, and the status and location of delivered or tasked resources.

Logistics/resource view Map by Esri; data courtesy of City of Naperville.

Public information view

The public information viewer is a web-enabled map viewer made available to the public. It can serve as a comprehensive local government citizen service application at all times, giving the public continuous access to civic operations. When a catastrophic event occurs, it can be shifted into emergency management mode to provide authoritative information notifying the public about current conditions and alerts, designated evacuation areas and assembly points, available medical and food and shelter facilities, and roadblocks and restricted zones. It also supports citizen requests for service and can post feeds uploaded by citizens recording events and conditions in the field via social media portals. This viewer is very targeted and lightweight, with limited tools and data for public access.

Public information view Map by Esri; data courtesy of City of Naperville.

Briefing templates

As witnessed a number of times throughout this book, briefing cycles in the midst of a catastrophic event can sometimes take up to 12 hours or longer to update. This lag time does not enable command and operations personnel to make effective time-sensitive emergency decisions. Over such an extended amount of time, conditions on the ground may differ greatly from conditions reported during the last operational period briefing. With the advent of mobile GIS, this cycle has been significantly reduced to almost real time, as field data is gathered and synchronized wirelessly with the geodatabase at the incident command post and EOC for rapid display and decision support. This enhanced capability updates all incident personnel as events change and provides current and timely information during regularly scheduled briefing sessions.

As incident briefings are conducted in the EOC, they are typically done following the NRF's ESF protocol of using text-based slides and marked printed maps. ArcGIS for Emergency Management has evolved this model to create dynamic maps that support briefings with separate live map views representing the ESFs that align with common briefing outlines. ArcGIS

Explorer Desktop and the web-based ArcGIS Explorer Online provide the presentation platform where briefing presentations anchored in the ArcGIS common operating platform provide a current and accurate representation of ongoing incident conditions. This evolution from static displays of on-site conditions at a fixed place and past time to real-time, fully integrated dynamic pathways through the data at any time in the life cycle of an event is a huge leap in enabling comprehensive situational awareness. Incident management personnel, duty officers, field personnel, and all event partners now have fully integrated live updates at any place and time that align with the NRF's targeted ESFs, enabling decision makers to identify the most critical priorities and allocate and assign appropriate resources.

During the Deepwater Horizon oil spill and the New Madrid National Level Exercise (NLE2011), mission-specific briefing templates were deployed to support various operations during multifaceted and complex operations. Command, operations, logistics, and incident management personnel were able to develop accurate and timely response and recovery plans as a result of having targeted viewers with near-real-time data integration.

During the NLE,[3] dynamic ESF views were implemented to support briefings and updates and to illustrate very specific infrastructure conditions and needs as the event unfolded.

NLE11 briefing map (Homeland Security [HLS] Summit 2011) Map by Esri; data courtesy of Food and Agriculture Organization of the United Nations, National Geophysical Data Center, US Geological Survey, NASA, Automotive Navigation Data (AND), TomTom, and US Environmental Protection Agency.

---

3 See the NLE video on YouTube (http://www.youtube.com/watch?v=eOBHn4G_JUw) for a comprehensive overview of the use of mission-specific map viewers embedded in the common operating platform during the NLE.

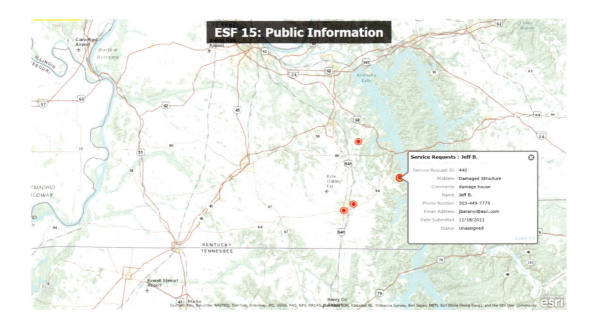

NLE11 briefing maps (Homeland Security [HLS] Summit 2011) Map by Esri; data courtesy of National Geospatial-Intelligence Agency, DeLorme, US Geological Survey, Eros Data Center, US Environmental Protection Agency, US Park Service, Automotive Navigation Data (AND), and TomTom.

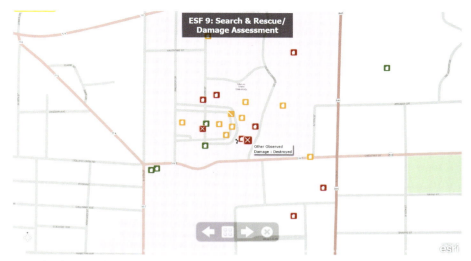

NLE11 briefing map (Homeland Security [HLS] Summit 2011) Map by Esri, data courtesy of TomTom.

The use of these mission-specific dynamic live briefing maps enabled all event partners to be fully informed of current conditions at the site of an incident as the event unfolded.

NLE11 mission-specific map viewer used at the NLE2011 briefing session in Kentucky
Courtesy of Kentucky Emergency Management.

Incident management staff and emergency operations personnel were then able to make critical decisions based on true conditions at the site, ultimately securing the safety of people and property more efficiently and effectively.

Summary

The level of activity and commotion surrounding a catastrophic event is almost always overwhelming. As our decision support systems evolve, emergency management personnel are better able to manage and control incident command and information workflows. A well-defined geospatial information system, embedded in a common operating platform, provides decision makers with this capability because it enables the greatest degree of situational awareness of an event before, during, and after it unfolds.

Chapters 1 through 5 described the critical components of this important information system. Data management is the most vital piece of this system. Without accurately knowing what and where incidents are occurring and how communities are affected by them, commanders and managers are essentially doing the best they can with limited information, which impedes their ability to achieve a successful collective response. Embedding this incident information in a dynamic geodatabase; performing analyses to understand specific vulnerabilities, potential impacts, and overall consequences; and then sharing this information among all event partners through mission-specific situational awareness map viewers enables effective response to events that threaten the safety of a community.

Having the capability to understand how events can affect a community and its citizens before they occur enables emergency planners to better prepare, prevent, and mitigate community loss. Using GIS to support vulnerability assessments modeling enables an understanding of accurate and realistic implications for the future security of a community. Observations and updates from field personnel go one step further to confirm real-time conditions that can be readily integrated into decision-making workflows as changes occur and as incidents unfold.

Channeling all of this information and analysis into a well-designed briefing workflow serves the ultimate purpose of informing those charged with incident management responsibilities. Providing the current incident status, potential incident growth, and required actions to meet priority incident objectives using live, real-time mission-specific briefing maps aligned with the NIMS and ESFs is the final piece that defines a successful emergency management information workflow. This configuration enables the greatest degree of situational awareness possible as an event is underway.

The next, and final, chapter of this book provides an implementation guide for each of the components of a GIS-based common operating platform. The combination of data needs, planning and analysis models, and field operations come together in this platform as a fully implemented and integrated decision support system for emergency management. This guide directs users through the implementation of templates found on the ArcGIS Resources Local Government—Public Safety Community website that model the best practices defining the most effective common operating GIS platform for comprehensive situational awareness.

Chapter 6: Implementation guide

Throughout this book we have discussed how GIS is applied to each of the major emergency management workflow patterns to enhance situational awareness before, during, and after emergency events. The book contains examples that show best practices across the industry in GIS data management, analysis, and field mobility. This chapter focuses on implementing the pieces of the ArcGIS common operating platform that deliver actionable information to managers and commanders. A selected set of templates and tools are explored that publish data embedded in the platform, bringing it to those making the critical decisions that save lives and protect property. We focus on the implementation and configuration of key situational awareness viewers and map products using sample data provided from Naperville, Illinois. Configuration notes addressing how they can be configured specifically for your jurisdiction are also presented to support the implementation of a system that enables the greatest situational awareness of emergency events in your own community.

Introduction

Before building a GIS in support of emergency management, careful consideration needs to be given to the problems to be solved and the answers most likely needed in all stages of an emergency operation. What questions will decision makers be asking of their GIS staff? The US wildfire community has done a great job articulating the map-based products they normally require as they face wildfires and all-hazards incidents. This community has developed a GIS Standard Operating Procedures on Incidents (GSTOP)[1] that not only defines key GIS components and terminology, such as minimum essential datasets, symbology, and directory structures, but also the standard map products that are expected of the GIS staff. Some of these products include the following:

- Incident Action Plan (IAP) Map
- Transportation Map
- Incident Briefing Map
- Progression Map
- Aerial Operations Map
- Damage Assessment Map
- Public Information Map

Although these products and information directly support an incident with immediate tactical products, rather than the strategic products emergency management agencies or EOCs may typically require, a lot can be learned from the wildfire community and the GSTOP.[2]

Another important document is the Federal Interagency Geospatial Concept of Operations (GeoCONOPS)[3] from the US Department of Homeland Security's Geospatial Management Office (GMO). This document lays out how the GIS community should support the federal government under the NRF,[4] specifically by organizing data and products along the lines of FEMA's fifteen ESFs. This organizational structure is critical to the success of any emergency operation because that is how the response community is organized. Often in the EOC there are tents labeled by ESF on top of many workstations. As discussed in chapter 5, ESF role-based viewers were one of the key outcomes of the 2011 NLE that simulated a large earthquake in the New Madrid Seismic Zone.[5]

---

1 GIS Standard Operating Procedures Project (GSTOP), National Wildfire Coordinating Group, Information Resource Management Working Team, Geospatial Task Group, June 2006, http://www.nwcg.gov/pms/pubs/GSTOP7.pdf.

2 The Bureau of Land Management offers a class, Geographic Information System Specialist for Incident Management Training (S-341). The first part of the class is available for free online at http://www.ntc.blm.gov/krc/uploads/223/GISS.html. The second portion of the class is a three-day instructor-led class.

3 "Federal Interagency Geospatial Concept of Operations (GeoCONOPS), version 3.0," US Department of Homeland Security, June 2011, http://www.napsgfoundation.org/attachments/article/113/DHS_Geospatial_CONOPS_v3.0_8.5x11.pdf.

4 "National Response Framework," Federal Emergency Management Agency, August 2012, http://www.fema.gov/national-response-framework.

5 Toward the end of the exercise week, the Esri team supported briefings in the Commonwealth of Kentucky's Emergency Operation Center via a live ArcGIS Explorer Online presentation with the data and information products organized by ESF. This was a very effective and efficient way to convey this information. Their work was presented at the 2011 Esri Homeland Security Summit and the video of that presentation is available here: http://proceedings.esri.com/library/userconf/hss11/videos/06-lg.html.

In support of these role-based viewers, we have organized the appropriate layers from the LGIM reviewed in chapter 2 by ICS and ESF role.[6] Regardless of which emergency support structure you choose, you must consider how to organize and relate all of the data, information, and tools available within the ArcGIS platform, or it will be simply be too overwhelming to meet the demands of the decision makers you are supporting. Understanding the workflow and decisions that need to be made will help you target your information products. Having a good understanding of the problems to be solved and the decisions to be made will allow you to provide better information so more effective decisions can be made that will ultimately save lives and secure property.

The documents and training available in the emergency management community will help your understanding, especially in an operational context when time constraints are compressed. Also note that training exercises, both large and small, are a great way to test your system and get feedback from your decision makers on how you have organized your information in a consequence-free environment. In addition to exercises, special events that activate the EOC are also good opportunities to practice emergency procedures and operational workflows.

ArcGIS Resources

Several years ago, Esri started an initiative, the ArcGIS Resource Center (now called ArcGIS Resources), to make it easier to deploy ArcGIS to support various industries. There is a Public Safety community on the ArcGIS Resources site with maps and applications specific for emergency management. In this chapter, we will review implementation considerations for a select set of templates available on this site. When you download the templates, you will find more detailed instructions in the supporting documentation than are provided in this overview.

The purpose of the templates is to make it easier to deploy ArcGIS within your organization to support a specific mission. We have collected best practices from the user community, adhered to national and international standards where appropriate, and incorporated feedback from the user community to provide these templates and tools. These applications are updated regularly and new applications are added all the time. Please be sure to keep an eye on the ArcGIS Resources

---

6 See appendix B for a cross-referenced table aligning the feature datasets of the LGIM with ICS and ESFs.

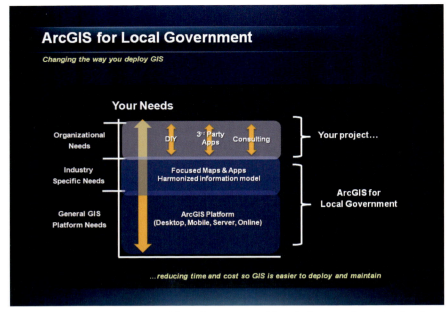

ArcGIS for Local Government—Public Safety: How to address industry and organizational needs with the best ArcGIS platform and products Esri.

Local Government—Public Safety website (http://resources.arcgis.com/en/communities/local-government/) for continuous updates.

At the time of this writing, we chose to focus on three templates available with the release of ArcGIS 10: Public Safety COP, Public Safety Damage Assessment, and Emergency Management Maps. These templates by no means replace Esri partner applications. In fact, many partner applications have taken this content and extended it, in some cases beyond the scope and capabilities of the templates themselves. They do, however, touch on three of the most common applications of GIS in the emergency management workflow. Additional templates are listed and briefly discussed at the end of the chapter.

The templates on the ArcGIS Resources Local Government—Public Safety website are updated regularly, just like Esri software. This book was written with the version of the templates available at ArcGIS 10. These templates will be updated as the software evolves, and the templates will be ported to the next release. For content specific to this book, a companion site has been set up at http://esripress.esri.com/bookresources, where you can find information on the most current version of the templates and updates to the information covered in this chapter.

Training and proficiency

To be successful with these templates, you'll need to be proficient with both ArcGIS for Desktop and ArcGIS for Server. Several training options exist to help you increase your proficiency. You will need to know the basics of map creation and manipulation and the details of geodatabase management, especially as you shift toward implementing this for your jurisdiction. The following Esri courses are available for ArcGIS for Desktop to increase your proficiency:

- Learning ArcGIS Desktop (web course)
- ArcGIS Desktop III: GIS Workflows and Analysis (instructor-led, two-day course)
- Managing Editing Workflows in a Multiuser Geodatabase (instructor-led, three-day course)

The skills required for the desktop are consistent with the Esri ArcGIS for Desktop Associate (EADA) Technical Certification.

For ArcGIS for Server, three Esri classes are recommended:

- Introduction to ArcGIS Server (instructor-led, two-day course)
- ArcGIS for Server: Sharing Content on the Web (instructor-led, two-day course)
- ArcGIS Server: Web Administration Using the Microsoft.NET Framework (instructor-led, three-day course)

The templates contain detailed instructions, but you will need to know how to install and work with the templates before using them for your jurisdiction. Details such as permissions, caching fundamentals, types of services, and so on, are all important to your success with the templates.

You will also need to know how to use ArcGIS Viewer for Flex, upon which the COP is built. Courses exist to help you increase your proficiency with the ArcGIS Viewer for Flex. For an overview of the application, a free Esri live training seminar—Introduction to the ArcGIS Viewer for Flex—is offered. As you look to extend the COP and build your own custom widgets, the Building Web Applications Using the ArcGIS API for Flex course will be helpful.

ArcGIS for Windows Mobile is a part of ArcGIS for Server used for the Damage Assessment template. The Authoring and Serving ArcGIS Mobile Projects instructor-led course will increase your mobile skills.

Details about these courses, as well as many other courses and web modules, are on the Esri Training website at http://training.esri.com.

Installation

Before exploring the templates, the Esri software needs to be properly installed and configured. Please be sure to check the system requirements for ArcGIS for Desktop,[7] ArcGIS for Server,[8] ArcGIS for Windows Mobile,[9] and ArcGIS for Smartphones and Tablets.[10]

In addition to having the software properly configured, it's also important to consider the load that your system can handle. Day-to-day operations may not incur a heavy load on your system, but when disaster strikes you'll want to be ready. The number of people involved in supporting the response to a large incident, and the thirst for current information, can seriously tax your system. Not only will you serve internal clients in the EOC, but you will also serve the public and media with information. Are you prepared to handle the load if your site makes it to a major news network website? It's important to keep in mind strategies for dealing with these issues.

Software architecture

Before installing and configuring the templates, the base ArcGIS system needs to be in place and tuned to optimal performance. The minimum requirements for successful implementation of the templates are ArcGIS for Desktop with the Maplex for ArcGIS extension (available with all levels of ArcGIS 10.1 for Desktop) and ArcGIS for Server Advanced Enterprise. Underlying a good installation is a good architecture. There are many considerations, more than we will cover in this book, but the following is a sample architecture that successfully runs both locally in a server environment and also in the cloud.

The following resources are helpful guides to designing a successful system architecture to support the ArcGIS platform:

- ArcGIS Resources Enterprise website: http://resources.arcgis.com/en/communities/enterprise-gis/
- System Design Strategies: http://wiki.gis.com/wiki/index.php/System_Design_Strategies
- System Design Strategies Esri instructor-led class: http://training.esri.com/gateway/index.cfm?fa=catalog.courseDetail&CourseID=50104470_10.X

---

7 "ArcGIS Desktop 10 System Requirements," http://resources.arcgis.com/content/arcgisdesktop/10.0/arcgis-desktop-system-requirements.
8 "ArcGIS Server 10 System Requirements," http://resources.arcgis.com/content/arcgisserver/10.0/arcgis-server-system-requirements.
9 "ArcGIS for Windows Mobile System Requirements," http://resources.arcgis.com/content/arcgis-mobile/10.0/system-requirements.
10 "ArcGIS for Smartphones and Tablets System Requirements," http://www.esri.com/software/arcgis/smartphones/system-requirements.html.

- *Building a GIS: System Architecture Design Strategies for Managers*, second edition, by Dave Peters (Esri Press, 2011)
- *Thinking About GIS: Geographic Information System Planning for Managers*, fourth edition, by Roger Tomlinson (Esri Press, 2011)

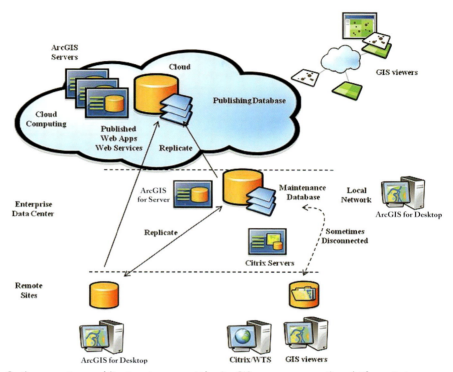

Optimum system architecture to support the ArcGIS common operating platform Esri.

Where are the templates?

To find and download the templates, visit the ArcGIS Resources Local Government—Public Safety website, http://resources.arcgis.com/en/communities/local-government/. Note that although the title of this community is "Local Government—Public Safety," the main and original content was focused primarily on emergency and disaster management. Over time, other templates have been added to support other aspects of public safety, including fire and law enforcement.

The Local Government—Public Safety content is located under the Communities tab. This takes you to the Local—Public Safety Community site. You can also get there directly by going to http://resources.arcgis.com/en/communities/local-government/01n40000000w000000.htm.

ArcGIS Resources website Esri; GeoEye satellite image. Copyright 2012. All rights reserved.

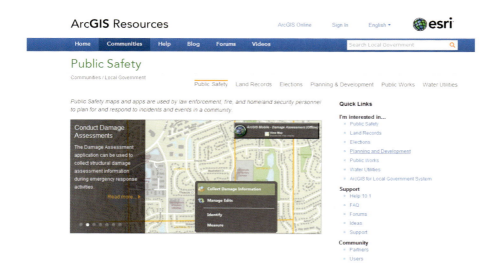

ArcGIS Resources Local Government—Public Safety Community site Esri.

Here you will find the various templates. Note that there are several versions of the applications going back to ArcGIS 9.3. Download the latest versions. In this section, we will focus on the *COP*, *Damage Assessment*, and *Maps Templates*, but you may also find other templates valuable. The templates used in this guide are configured with ArcGIS Viewer for Flex. Other options for situational awareness viewers are available as well, but no templates exist for those. The ArcGIS Viewer for Silverlight is a suitable option, especially if your organization has access to Microsoft/Silverlight developers. Additionally, you can use the ArcGIS for SharePoint to quickly configure the same applications.

You will also see other resources at the bottom of this web page. In addition to the templates, there is also the Public Safety blog (http://blogs.esri.com/esri/arcgis/category/subject-public-safety/) and recent tweets from the Esri Public Safety Twitter account (@GISPublicSafety), as well as videos on how to get started with the applications.

As you navigate through each template, notice the instructional notes guiding you through the installation and configuration of each component. Additional *configuration notes* follow the template guidelines, addressing extended capabilities for configuring the map viewers for use with local data. These remarks are intended to support the installation of a fully configured common operating GIS platform for emergency management planning and operations.

What's in the templates?

When you download a template and unzip the contents to your own local system, you will find four main components at the root level:

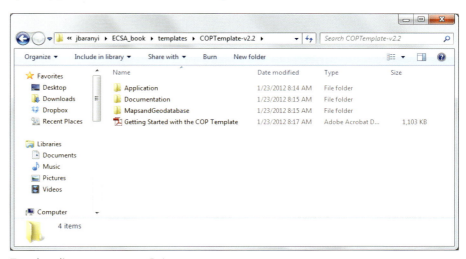

Template directory structure Esri.

- *"Getting Started with the COP Template"*—This PDF document contains the basic instructions for installing and configuring the template with the sample data and applications provided, as well as system and version requirements, basic configuration steps, and any associated release notes.
- *Application directory*—This folder contains sample configuration files that are edited in the configuration and customization process, add-ins, or other code custom to the particular template.
- *Documentation directory*—This folder contains more detailed information on the geodatabase elements. These are generally HTML documents exported from the GDB XRay tool.
- *Maps and Geodatabase directory*—This folder contains a sample Naperville, Illinois, geodatabase embedded in the LGIM, along with associated map documents. This sample LGIM dataset is downloaded with all of the selected templates reviewed in this guide.

The Public Safety COP template

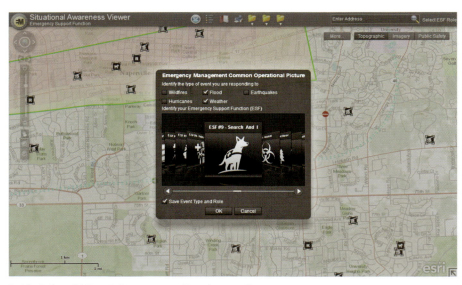

Public Safety COP   Esri; data courtesy City of Naperville.

The Public Safety COP template uses the ArcGIS Viewer for Flex viewer. It is the most current viewer and has been around the longest. It has been adopted by many emergency management agencies across the globe. It also has the largest set of available tools developed for it. Download the Public Safety COP template from http://esriurl.com/EMCOP.

The instructions for configuring the COP template on your local server are provided in the "Getting Started with the COP Template" documentation. Following is a quick-start guide through those steps.

Configuring the COP template on your server

The first step of configuring the COP template on your server is to download the Aggregated Live Feeds tools and scripts from the Aggregated Live Feed Community group on ArcGIS.com (http://www.arcgis.com/home/group.html?owner=Esri_Technical_Marketing&title=Aggregated Live Feed Community). These scripts will be used to keep the closest weather station geoprocessing tool updated with current wind direction information for the ERG tool.

The next step is to publish the basemap services found in the \COPTemplate\MapsandGeodatabase folder:

- ◆ ImageReferenceOverlay.mxd
- ◆ PublicSafety.mxd
- ◆ Topographic.mxd

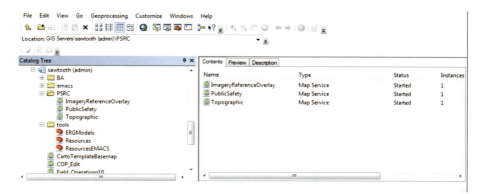

Publishing the basemap services Esri.

The Public Safety basemap used in the COP template is a derivative of the Local Government basemap, but provides more emphasis on public safety and critical infrastructure features. It contains more detail than other commercial or federal datasets, and forms the foundation of an effective public safety situational awareness map viewer (see map on page 144).

After publishing the basemap services, build the cache for each of the services. Building the cache will enable fast and efficient map display at all scales, ensuring that time-sensitive information is always on hand when critical decisions are being made.

COP using the Public Safety basemap Map by Esri; data courtesy City of Naperville.

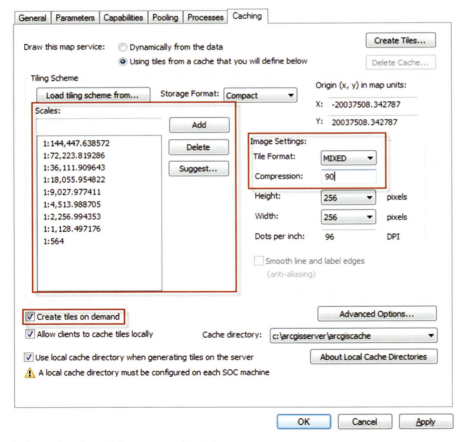

Cache settings for each basemap service Esri.

Because this is relatively large-scale data of a small area, delete any scales smaller than 1:144,447. Additionally, given the level of detail available in the data, add a scale of 1:564. For the Tile Format, select MIXED and increase the compression to 90 to save disk space. The option to create tiles on demand is recommended so you don't have to build all levels at once, and the users of the service can help build up the cache. Once you have all of these choices set, create the tiles and choose to build the tiles in the next dialog box.

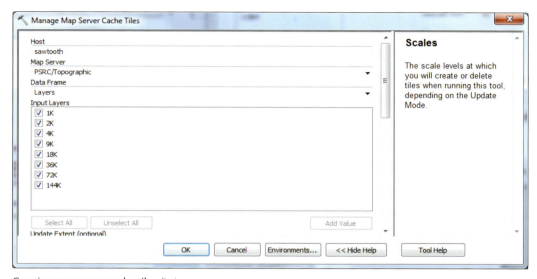

Creating map server cache tiles Esri.

Repeat this process to complete the cache building for all three map services.

After publishing the basemaps services, the next step is to configure the Aggregated Live Feeds scripts by updating the appropriate path information as well as updating the source of the live feed layer in the EmergencyOperations.mxd.

The next step is to publish the feature services. This includes the operational layers that will be edited as emergency events are planned for and unfold. First, create a LocalGovernment geodatabase inside ArcSDE Workgroup (or Enterprise) and copy in the EmergencyOperations and PublicSafetyPlanning feature datasets from the sample data provided. Grant appropriate permissions to the ArcGIS SOC user before publishing the map services.

Next, you need to install the ERG widget. A widget is a program that adds functionality to the map viewer.[11] In addition to widgets that

---

11 "Widgets in the ArcGIS Viewer for Flex," http://help.arcgis.com/en/webapps/flexviewer/help/index.html#/Widgets_in_the_ArcGIS_Viewer_for_Flex/01m30000001v000000/.

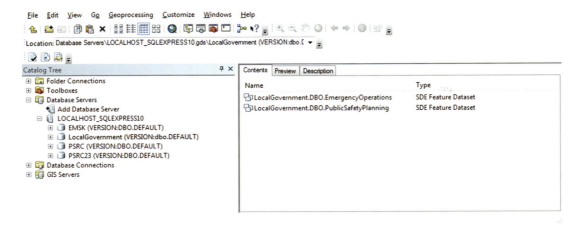

Publishing the feature services Esri.

perform basic functions—such as edit, search, sketch, and print—the Public Safety COP template embeds the ERG widget, as discussed in chapters 1 and 2, into the COP. This widget processes hazardous material data to delineate safety zones around a spill site.

The ERG widget relies on an underlying custom geoprocessing tool that is to be installed via the included installation package found at \\Application\ERGToolsforArcGIS10. Once this widget is installed, the ERGModels toolbox is published as a geoprocessing service.

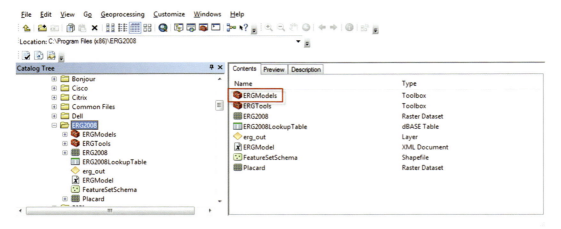

The ERGModels toolbox published as a geoprocessing service Esri.

After publishing the services, this geoprocessing service, along with the map services published in the previous steps, appear in the ArcGIS Services Directory.

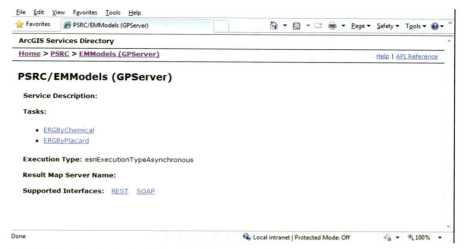

Published services in the ArcGIS Services Directory Esri.

The next geoprocessing service to publish is the NearestWeatherStation toolbox. Before you can publish it you need to edit the model and update the location of the Current_Metar feature class to the local copy that is being updated by the Aggregated Live Feeds methodology you set up previously.

Now that the basemap services, operational layer services, and geoprocessing services are published to your local server, the Public Safety COP application is ready to be configured and deployed.

Configuring and deploying the Public Safety COP application

The first step of configuring and deploying the Public Safety COP application is to use a text or XML editor to update the URLs within the main config.xml file found in the Applications folder. Point the three basemap services to your local server.

```
<widget left="0"    top="0"    config="widgets/HeaderController/EM_HeaderControllerWidget.xml" url="widgets/HeaderController/HeaderControll
<widget left="3" bottom="3" config="widgets/Coordinate/CoordinateWidget.xml" url="widgets/Coordinate/CoordinateWidget.swf"/>

<map wraparound180="true" initialextent="-9816800 5121400 -9805200 5128100" fullextent="-9816800 5121400 -9805200 5128100" top="40">
    <basemaps>
        <layer label="Topographic"    type="tiled" visible="false" alpha="1"
                 url="http://server.arcgisonline.com/ArcGIS/rest/services/World_Topo_Map/MapServer"/>
        <layer label="Topographic" type="tiled" visible="false" alpha="1"
                 url="http://yourserver/ArcGIS/rest/services/Topographic/MapServer"/>
        <layer label="Imagery" type="tiled" visible="false" alpha="1"
                 url="http://server.arcgisonline.com/ArcGIS/rest/services/World_Imagery/MapServer"/>
        <layer label="Imagery" type="tiled" visible="false" alpha="1"
                 url="http://yourserver/ArcGIS/rest/services/ImageryReferenceOverlay/MapServer"/>
        <layer label="Public Safety"    type="tiled" visible="false" alpha="1"
                 url="http://server.arcgisonline.com/ArcGIS/rest/services/World_Topo_Map/MapServer"/>
        <layer label="Public Safety" type="tiled" visible="true"  alpha="1"
                 url="http://yourserver/ArcGIS/rest/services/PublicSafety/MapServer"/>
    </basemaps>
    <operationallayers>
        <layer label="RIDGE Radar(NOAA Feed)" type="dynamic" visible="false" alpha="0.75"
                 autorefresh="60" url="http://gis.srh.noaa.gov/ArcGIS/rest/services/RIDGERadar/MapServer" />
```

Editing the config.xml file to point the basemap services to the local server Esri.

Update the URLs of the operational layer services as well.

```
<layer label="PDC Active Hazards(GHIN Feed)" type="dynamic" visible="false" alpha="1.0"
     autorefresh="60" url="http://ags.pdc.org/rest/services/GHIN/PDC_Active_Hazards/MapServer">
     <sublayer id="0" popupconfig="popups/PopUp_EMPDC.xml"/> </layer>
<layer label="NHSS Natural Hazards(USGS Feed)" type="dynamic" visible="false" alpha="0.75"
     autorefresh="60" url="http://rmgsc.cr.usgs.gov/ArcGIS/rest/services/nhss_haz/MapServer" />

<layer label="Emergency Operations" type="dynamic" visible="true" alpha="0.5"
     url="http://yourserver/ArcGIS/rest/services/EmergencyOperations/MapServer"/>

<layer label="Incident Point" type="feature" visible="true"
     popupconfig="popups/PopUp_EMIncidentPoints.xml"
     url="http://yourserver/ArcGIS/rest/services/PublicSafety/IncidentCommand/FeatureServer/1"/>

<layer label="Incident Line" type="feature" visible="true"
     popupconfig="popups/PopUp_EMIncidentLines.xml"
     url="http://yourserver/ArcGIS/rest/services/IncidentCommand/FeatureServer/2"/>

<layer label="Incident Area" type="feature" visible="true"
     popupconfig="popups/PopUp_EMIncidentArea.xml"
     url="http://yourserver/ArcGIS/rest/services/IncidentCommand/FeatureServer/3"/>

<layer label="Shelters" type="feature" visible="true"
     popupconfig="popups/PopUp_EMShelters.xml"
     url="http://yourserver/ArcGIS/rest/services/IncidentCommand/FeatureServer/7"/>

<layer label="Pet Collection Areas" type="feature" visible="true"
     url="http://yourserver/ArcGIS/rest/services/IncidentCommand/FeatureServer/8"/>

<layer label="Public Safety Resources" type="feature" visible="true"
     popupconfig="popups/PopUp_EMResources.xml"
     url="http://yourserver/ArcGIS/rest/services/IncidentCommand/FeatureServer/9"/>

<layer label="Resource Assignments" type="feature" visible="true"
     url="http://yourserver/ArcGIS/rest/services/IncidentCommand/FeatureServer/10"/>

<layer label="Road Blocks" type="feature" visible="true"
     popupconfig="popups/PopUp_EMRoadBlocks.xml"
     url="http://yourserver/ArcGIS/rest/services/IncidentCommand/FeatureServer/5"/>

<layer label="Access Points" type="feature" visible="true"
     url="http://yourserver/ArcGIS/rest/services/IncidentCommand/FeatureServer/6"/>

<layer label="US National Grid - 10Km" type="feature" visible="false"
     url="http://yourserver/ArcGIS/rest/services/IncidentCommand/FeatureServer/12"/>

<layer label="US National Grid - 1Km" type="feature" visible="false"
     url="http://yourserver/ArcGIS/rest/services/IncidentCommand/FeatureServer/13"/>
</operationallayers>
```

Editing the config.xml file to point the operational layer services to the local server Esri.

Next, configure the ERG widget (\Application\PublicSafetyCOP\widgets\ ERGChemical\ERGChemical.xml) and Report by Exception widget (\Application\PublicSafetyCOP\widgets\ReportByException\EM_GeoRSS_ ReportByException.xml) to point to your local server.

Configuration of the Public Safety COP is complete. It is now ready to be deployed.

After updating the main config.xml file for the application you'll also have to update the configuration files for the other Emergency Support Function config.xml files. The easiest way is to simply perform a global search and replace on those configuration files.

Navigate to http://<your server>/PublicSafetyCOP/index.html to open the COP map viewer.

The Public Safety COP has been configured to be served from your local server. It is designed to display topographic, imagery, or the public safety

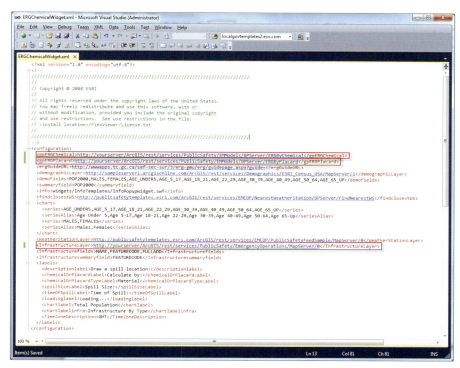

Editing the config.xml file to point the geoprocessing widgets to the local server Esri.

The Public Safety COP with configured basemap and operational layers and geoprocessing widgets Esri; map data courtesy of City of Naperville.

basemaps. It supports a collection of widgets docked along the top tool-bar, including the ERG geoprocessing tool configured in the earlier step. Additional customizations can be done to include other basemaps, operational layers, and widgets as required for a particular location or event or stream-lined for a mission-specific emergency support function. Further configuration techniques and suggestions are included at the end of this chapter.

The Damage Assessment template

A lot of information needs to be collected after an emergency event. Oftentimes, recovery personnel are quick to have paper forms ready for the information to be collected by field teams. As discussed in chapter 4, the manual process of recording damage assessment and recovery information can be quite time consuming and subject to error. It is important, however, not to dismiss these paper forms completely, because they provide the primary data structure defining what information is needed to record the extent of damage in a community after an emergency event.

The Public Safety Damage Assessment template is based on typical state damage assessment forms for collecting damage assessment data. It includes forms for Individual Assistance—Home, Individual Assistance—Business, and Public Assistance. The feature classes defined in your geodatabase can be modeled after this standard, although you may need to modify it accord-ing to the specific needs of your organization (see figure on page 151).

Download the Public Safety Damage Assessment template for ArcGIS 10 from http://esriurl.com/EMDamageAssessment.

This template contains sample field collection forms and basemap data for Naperville, Illinois. In addition to the minimum software configuration, including ArcGIS for Desktop with the Maplex extension and ArcGIS for Server Advanced Enterprise, this template also requires ArcGIS for Windows Mobile.

The instructions for configuring the Damage Assessment template locally are provided in the "Getting Started" documentation. Following is a quick-start guide through those steps.

Configuring the Damage Assessment template

Two options are available for using the template: (1) a fully disconnected mode where field personnel work offline while collecting data, and then synchronize the data back to the enterprise geodatabase after the data col-lection is complete, and (2) a connected or occasionally connected environ-ment where the data is synchronized and updated in real or near-real time. The following overview focuses on the latter configuration.

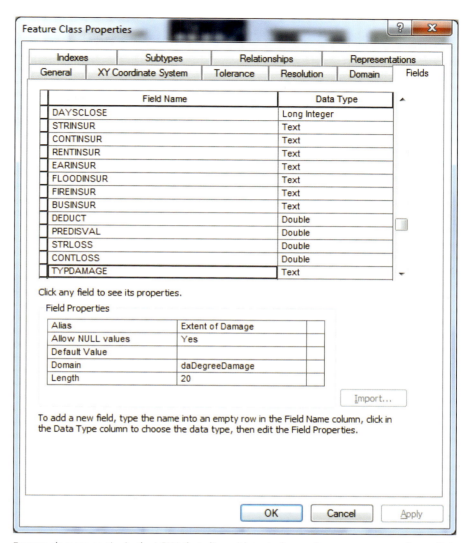

Feature class properties in the LGIM that align with typical state damage assessment forms Esri.

The first step in configuring the Damage Assessment template is to publish the daytime and nighttime mobile basemap services (MobileDay Basemap.msd, MobileNightBasemap.msd) (see page 152, top).

The second step is to build the mobile cache. The type of cache that you create to improve performance can be leveraged by ArcGIS for Windows Mobile so you can work more efficiently in a disconnected or occasionally connected environment. The cache settings are similar to those used for the basemap cache for the COP template (see page 152, bottom).

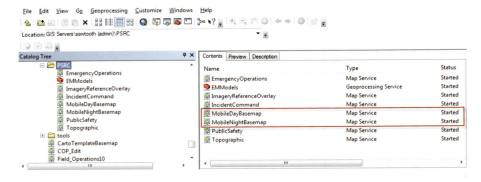

Publishing the mobile basemap services Esri.

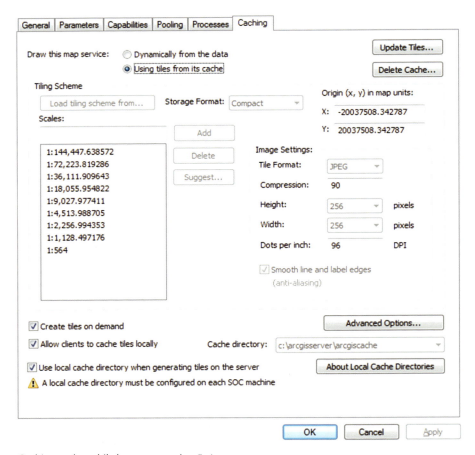

Caching each mobile basemap service Esri.

Next, publish the mobile map service that will be used in a connected environment. This map service has the features that will be edited in the field and synchronized back to the server when connectivity is established.

Open the DamageAssessment.mxd found in the Maps and Geodatabase directory and repoint the data sources to the ArcSDE geodatabase that you set up when configuring the COP—namely the DamageAssessment and USNationalGrid feature classes within the EmergencyOperations feature dataset. When publishing the new service, make sure that the Mobile Data Access option is selected.

Select Mobile Data Access Esri.

The map service is now ready to receive data from the field. It requires a damage assessment application on a mobile device to be configured and deployed to collect the necessary information (see page 154, top).

Configuring and deploying the Damage Assessment application

A mobile damage assessment application, DamageAsessmentOnline.wmpk, is already provided in the template, located in the Application folder. This application was created using Mobile Project Center (http://resources.arcgis.com/en/help/windows-mobile/app/index. html#///007v00000009000000), a feature of ArcGIS for Windows Mobile.

Copy the mobile damage assessment application and basemaps to directories that are accessible and registered with the ArcGIS for Windows Mobile application (see page 154, bottom).

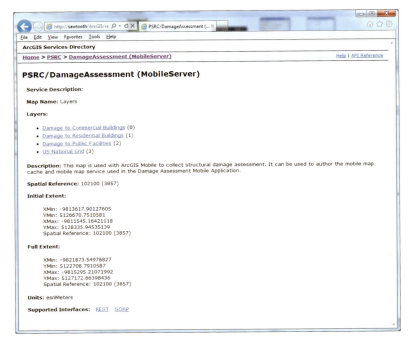

Published services in the ArcGIS Services directory Esri.

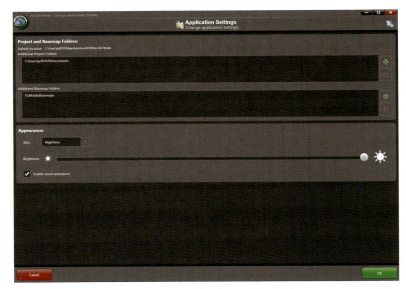

ArcGIS Mobile Damage Assessment application settings Esri.

Edit the DamageAssessmentOnline.amp file in the P_DamageAssessment Online directory, and edit the mobileServiceURL to point to the URL of your mobile map service (note this is the SOAP end point, not REST).

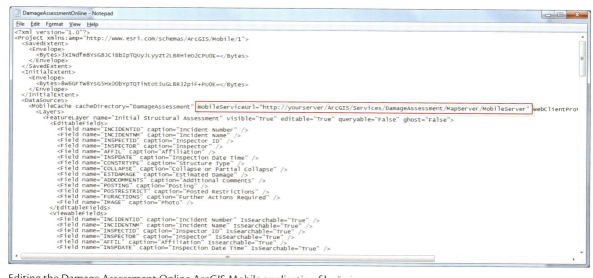

Editing the Damage Assessment Online ArcGIS Mobile application file Esri.

The data entry application is ready to be deployed.

ArcGIS Mobile Damage Assessment Online data entry application Esri.

The damage assessment information can now be collected day or night in the field using a laptop or mobile device and synchronized back to the geodatabase in real or near-real time.

Daytime mobile damage assessment map By Esri; data courtesy of City of Naperville.

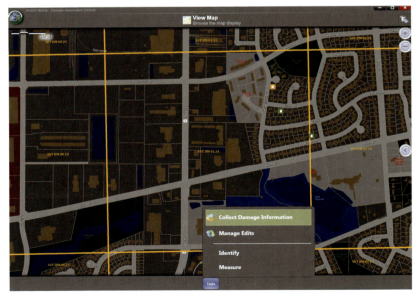

Nighttime mobile damage assessment map By Esri; data courtesy of City of Naperville.

Maps template

Although the ArcGIS common operating platform relies heavily on digital map products, there is still an overwhelming need for paper maps to support the complete emergency management workflow. Map posters continue to be the traditional method of providing situational awareness during a disaster. Templates for standard map products are critically important to generate these maps quickly.

The Emergency Management Maps template can be downloaded from http://esriurl.com/EMMaps.

The template provides a selected set of the standard map products from the GSTOP. Although the previous templates were embedded in an ArcGIS for Flex environment and delivered via a web browser, this template is prepared for the ArcGIS for Desktop platform. The template includes map layouts for the following critical map products that are heavily depended upon as an emergency event unfolds:

+ Hazard Map
+ Incident Action Plan
+ Incident Briefing Map

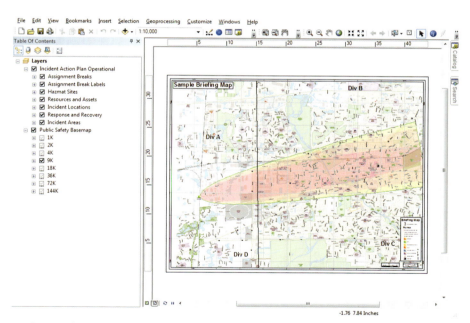

Incident Briefing Map template By Esri; data courtesy of City of Naperville.

The template also includes an example Incident Action Plan Map Book created using Data Driven Pages.[12] The "Getting Started" documentation for the Maps template provides detailed instructions on configuring these templates. There is an ArcGIS style file and two fonts to install. Additionally, some configuration settings for the Data Driven Pages will need to be updated. Finally, there is a Python script provided to assemble the Incident Action Plan Map Book PDF.

The logo, graticule preferences, and production information are all customizable to any local jurisdiction. Additionally, Esri Production Mapping (http://www.esri.com/software/arcgis/extensions/production-mapping/index.html) users can access a modified version of this template for that environment as well. The Production Mapping for Emergency Management Maps template can be downloaded from http://esriurl.com/EMMapsProd.

Other Local Government—Public Safety Community templates

The templates configured in this chapter provide the foundation for delivering actionable information from the ArcGIS common operating platform to those making critical decisions in the event of an emergency. A collection of other templates on the ArcGIS Resources Local Government—Public Safety Community site can extend the system even further, as well as provide specialized functions required during mission-specific operations.

The Public Safety Special Event Planning template provides a data model for special events that may occur in your jurisdiction. It is designed to configure the information your agency collects and shares with the public in preparation of a planned event. It is helpful to embed this information in the COP, as well as to produce paper PDF maps for distribution. The Public Safety Special Event Planning template can be downloaded from http://www.arcgis.com/home/item.html?id=d80b88a7acf841249bdacc259a2c7c94.

A video with detailed visual instructions on how to use this template is also provided at http://resources.arcgis.com/gallery/video/local-government/details?entryID=70CCB322-1422-2418-A00B-4B2A6BD39FA3.

The Citizen Service Request template provides a channel for the public to report information on damaged structures and local infrastructure. This is an example of how to leverage crowd-sourced or volunteered geographic information, as discussed in chapter 2. The Citizen Service Request template can be downloaded from http://esriurl.com/EMCitizenEngagement.

---

12 "What Are Data Driven Pages?" http://resources.arcgis.com/en/help/main/10.1/index.html#//00s90000003m000000.

Public Safety Special Event Planning template By Esri; data courtesy of City of Naperville.

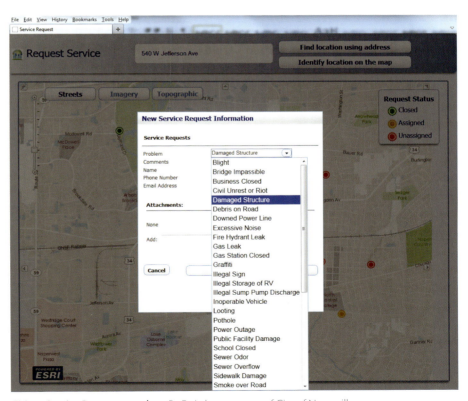

Citizen Service Request template By Esri; data courtesy of City of Naperville.

Configuration notes

Many components in the data models and map viewers support additional configuration but do not require programming development. Following is a discussion of possible configurations that can enhance the ArcGIS for Emergency Management platform with the look and feel of your local jurisdiction.

LGIM

As noted earlier, all of the templates are delivered already populated with the sample dataset of Naperville, Illinois. To use the model for your own local data, an empty schema package can be downloaded from http://www.arcgis.com/home/item.html?id=5f799e6d23d94e25b5aaaf2a58e63fb1.

Several different options are available for loading data into this data model. The ArcGIS for Desktop Append[13] tool allows for field mapping when the schema type is set to NO_TEST. For a more complex migration of data, Extract, Transform, and Load (ETL) tools may be required, which can be created and run using the ArcGIS Data Interoperability[14] extension. At the time of this writing, template ETL tools for the data model are in development and will be posted on the ArcGIS Resources Local Government—Public Safety Community site when available. Detailed instructions on getting started with this schema package are posted on the *Implementing the Local Government Information Model with ArcGIS 10* blog.[15]

Note that the data for Naperville, Illinois, and the schema-only layer package use the Illinois state plane map projection. When customizing the LGIM, the coordinate system needs to be reprojected to one appropriate for your local jurisdiction.

Configuring the ArcGIS Viewer for Flex map

Changing the "brand" of the application to include your organization's logo and presentation graphics gives the map viewer the unique look and feel of your local jurisdiction. Following are ways you can customize the look of your application:

- The \<title\> and \<subtitle\> tags near the top of the main config.xml control the text of the application.

---

13 "Append (Data Management)," http://resources.arcgis.com/en/help/main/10.1/index.html#//001700000050000000.

14 "ArcGIS Data Interoperability: Data Transformation/Spatial Extract, Transform, and Load," http://www.esri.com/software/arcgis/extensions/datainteroperability/spatial-etl.html.

15 *Implementing the Local Government Information Model with ArcGIS 10,* http://blogs.esri.com/Dev/blogs/localgovernment/archive/2010/07/30/Implementing-the-Local-Government-Information-Model-with-ArcGIS-10.aspx.

- The <logo> tag points to the path of the image used in the upper left. Note that the optimum size of the image is 48 by 48 pixels.
- The <style> tag and associated subtags (<colors>, <alpha>, <font>, and <titlefont>) are used to control the visual appearance of the application. See the help topic "Setting Styles for the Viewer"[16] for additional details.

```
1<?xml version="1.0" ?>
2<!--
3////////////////////////////////////////////////////////////////////////////
4//
5// Read more about ArcGIS Viewer for Flex - http://links.esri.com/flexviewer
6//
7////////////////////////////////////////////////////////////////////////////
8-->
9<configuration>
10     <title>Enabling Comprehensive Situational Awareness</title>
11     <subtitle>Powered by ArcGIS Server</subtitle>
12     <logo>assets/images/em_logo.png</logo>
13     <style>
14         <colors>0xFFFFFF,0x333333,0x101010,0x000000,0xEAC65E</colors>
15         <alpha>0.8</alpha>
16     </style>
17     <!-- replace the following url with your own geometryservice -->
```

Editing the title, subtitle, logo, and style tags Esri.

Customized map viewer banner Esri.

- The initial and full extent of the map window are both controlled by the initialextent and fullextent attributes of the <map> tag of the config.xml. A handy utility called the Extent Helper allows users to zoom in to an area of interest and cut and paste the extent for use both by the <map> tag and the Bookmark widget. The utility is found at http://help.arcgis.com/en/webapps/flexviewer/extenthelper/flexviewer_extenthelper.html.

```
26    <widget left="0"  top="0"     config="widgets/HeaderController/HeaderControllerWidget.xml" url="widgets/HeaderController/HeaderController
27
28    <map wraparound180="true" initialextent="-9816800 5121400 -9805200 5128100" fullextent="-20000000 -20000000 20000000 20000000" top="40">
29       <basemaps>
30          <layer label="Streets" type="tiled" visible="true"
31              url="http://server.arcgisonline.com/ArcGIS/rest/services/World_Street_Map/MapServer"/>
32          <layer label="Aerial"  type="tiled" visible="false"
33              url="http://server.arcgisonline.com/ArcGIS/rest/services/World_Imagery/MapServer"/>
```

Editing the initial and full map extent windows Esri.

16 "ArcGIS Viewer for Flex," http://resources.arcgis.com/en/help/flex-viewer/concepts/index.html#/Setting_styles_for_the_Viewer/01m30000000s000000/.

◆ The BookmarkWidget.xml file allows customization of the bookmark extents so that they are appropriate for your local data and mission-specific map views. To configure the bookmarks, edit the file in the widgets\Bookmark directory with extents more appropriate for your jurisdiction.

Editing the bookmark extents and names Esri.

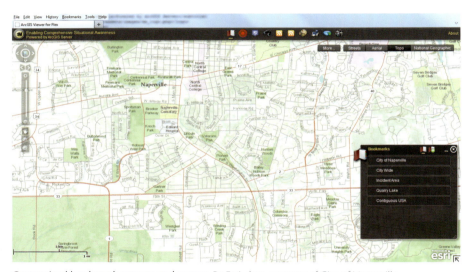

Customized bookmark extents and names By Esri; data courtesy of City of Naperville.

Basemaps

Many well-designed and effective general-purpose basemaps are available from ArcGIS Online, such as Streets, Topographic, and Imagery, to name a few. Additionally, some special-purpose layers, such as Light Gray Canvas, National Geographic, World Navigation Charts, Ocean Basemap, and OpenStreetMap are now available to quickly configure within the COP. You can also blend your local data in the LGIM with national and international basemap services to supplement the map with detailed local features. All of this is done by adding the appropriate layer tags within the basemap section of the main config.xml file.

```xml
<map wraparound180="true" initialextent="-9816800 5121400 -9805200 5128100" fullextent="-20000000 -20000000 20000000 20000000" top="40">
  <basemaps>
    <layer label="LOCAL StreetMap"      type="tiled" visible="false" alpha="1"
          url="http://localhost/ArcGIS/rest/services/Navteq_NA_WM/MapServer"/>
    <layer label="Topographic"     type="tiled" visible="false" alpha="1"
          url="http://server.arcgisonline.com/ArcGIS/rest/services/World_Topo_Map/MapServer"/>
    <layer label="Topographic" type="tiled" visible="false" alpha="1"
          url="http://localgovtemplates.esri.com/ArcGIS/rest/services/Topographic/MapServer"/>
    <layer label="Imagery" type="tiled" visible="false" alpha="1"
          url="http://server.arcgisonline.com/ArcGIS/rest/services/World_Imagery/MapServer"/>
    <layer label="Imagery" type="tiled" visible="false" alpha="1"
          url="http://localgovtemplates.esri.com/ArcGIS/rest/services/ImageryReferenceOverlay/MapServer"/>
    <layer label="Public Safety"     type="tiled" visible="false" alpha="1"
          url="http://server.arcgisonline.com/ArcGIS/rest/services/World_Topo_Map/MapServer"/>
    <layer label="Public Safety" type="tiled" visible="true"  alpha="1"
          url="http://localgovtemplates.esri.com/ArcGIS/rest/services/PublicSafety/MapServer"/>
    <layer label="National Geographic"    type="tiled" visible="false"
          url="http://services.arcgisonline.com/ArcGIS/rest/services/NatGeo_World_Map/MapServer"/>
    <layer label="Nav Charts"     type="tiled" visible="false"
          url="http://services.arcgisonline.com/ArcGIS/rest/services/Specialty/World_Navigation_Charts/MapServer"/>
    <layer label="Gray Canvas"     type="tiled" visible="false"
          url="http://services.arcgisonline.com/ArcGIS/rest/services/Canvas/World_Light_Gray_Base/MapServer"/>
    <layer label="OpenStreetmap" type="osm" visible="false" />
  </basemaps>
  <operationallayers>
    <layer label="NOAA Weather and Watches" type="dynamic" visible="false" alpha="0.7"
```

Editing the layer tags in the basemap section of the main config.xml file Esri.

When depending on online basemaps, it is important to consider having a fully cached copy in-house in case the system goes offline and needs to run disconnected from the Internet for a period of time. Caching local data can also be done from data included on the Data and Maps for ArcGIS DVDs from Esri or the purchased StreetMap Premium for ArcGIS data. Many federal and state users in the United States have access to these data layers from the National Geospatial-Intelligence Agency's HSIP dataset. More advanced options include getting Data Appliance for ArcGIS to completely guarantee availability of basemap data regardless of connectivity.

Live operational data feeds

The COP template is already configured with a sampling of live data feeds. Including others that are local or regional will enhance situational awareness for your organization. A variety of available live operational data feeds from outside public organizations can be readily consumed within the Public Safety COP. Examples include the USGS, National Health Security Strategy, and Pacific Disaster Center (PDC) feeds.

Other live feeds of information from partner enterprise systems and agency operations, such as CAD, CIMS, or US Department of Transportation traffic and road conditions, enhances the overall performance of ArcGIS common operating platform. Several mapping techniques, such as event and query layers, add these data features to the COP.

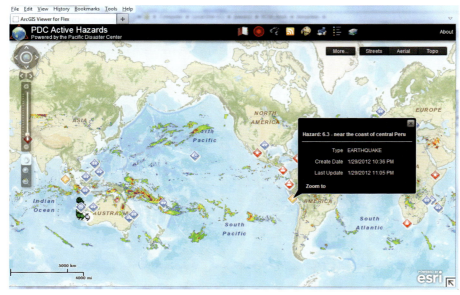

Live data feeds of active hazards in the Pacific Disaster Center Map viewer Map by Pacific
Disaster Center (PDC); data courtesy of Intact Forest Landscapes (IFL)—Potapov et al. 2008.

Widgets and geoprocessing services

A wide variety of widgets are available for integration into the COP. Each
supports a specific function, from creating simple bookmarks, legends,
and charts, to more complex geoprocessing widgets that perform queries
and searches, locate addresses, or extract subsets of data from a database.
The core application includes many well-designed ones, such as the Chart
and Locator widgets, and the COP template adds a few more, including
the ERG, Report by Exception, and Find Closest Facility widgets. Many
widgets have also been developed by the user community, which you can
browse and search by clicking the Community tab at http://www.esri.com/
flexviewer. Some of these widgets are at the current ArcGIS Viewer for Flex
10 version, and other older ones may need to be recompiled to the same
version you are using. For the best use in emergency management planning
and operations, see the table in appendix C that shows how the available
widgets align to each ICS and ESF role.

The ArcGIS Viewer for Flex now supports geoprocessing directly. You
are no longer required to do any customization to consume geoprocessing
services from ArcGIS for Server. This means than any of the analytic
workflows that you may have traditionally performed on the desktop can
be enabled by the server and quickly consumed within the Flex viewer.
In other words, the administrators of the ArcGIS System can grant

access to that analytic power to a larger audience. The samples provided by the ArcGIS Viewer for Flex include a set of great examples,[17] such as the Message in a Bottle, Population Summary, Drive Times, Calculate Viewshed, Extract Data, and Surface Profile geoprocesses.

Sample geoprocessing services available in the ArcGIS Viewer for Flex viewer Data courtesy of USGS Eros Data Center and TomTom.

Example of the Drive Time geoprocessing service included in the ArcGIS Viewer for Flex viewer Data courtesy of USGS Eros Data Center and TomTom.

---

17 "ArcGIS Viewer for Flex," http://help.arcgis.com/en/webapps/flexviewer/live/index.html?config= apps/config-geoprocessing.xml.

Field mobility

Field mobility options are also configurable to better suit to an organization's mission-specific on-site needs:

- ◆ *Field Crew Task*—Being able to track the location of teams in the field in real time is a compelling aspect of ArcGIS Mobile. The Field Crew Task[18] has the convention to automatically log the position of a GPS-enabled field device to a set of feature classes. The LGIM already contains the schema for the Field Crew Task. Publishing the Current Location feature class as a map service adds this feature to the COP, enabling real-time tracking of crew locations. Knowing the current location of field personnel is not only helpful from a safety perspective, but also helps managers and incident commanders direct the work of the on-site team in real time, allocating resources as needed to meet the challenges faced by response and recovery crews as an emergency event unfolds.

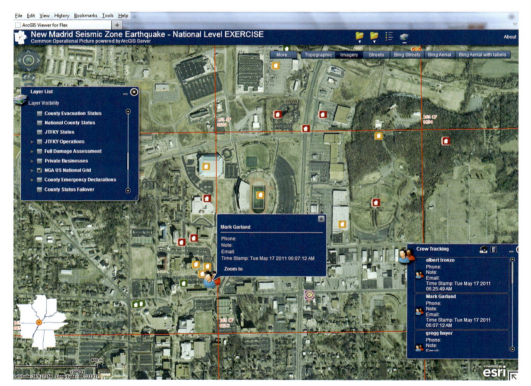

Use of the Field Crew Task service during NLE2011 Data courtesy of NASA.

---

18 "ArcGIS Resource Center," http://help.arcgis.com/en/arcgismobile/10.0/help/index.html#/
Configuring_the_View_Field_Crew_Task/007v00000018000000/.

◆ *Editing US National Grid*—Another useful field mobility configuration capability is included in the Damage Assessment template discussed earlier. Editing the values of US National Grid cells from *Searched* to *Not Searched* to track the status of a response and recovery team communicates progress of a task back to the COP, and to other personnel in other locations out in the field. The 1,000-meter US National Grid cells are available for all UTM zones by download as layer packages from ArcGIS Online.

◆ *Mobile COP*—Smartphones and tablets are also clients to the ArcGIS system. Using the ArcGIS.com application for the iOS, Windows Phone 7, or Android, anyone can view the intelligent web maps that were created by the user community and organized in the ArcGIS Online map gallery.

Information can get out to the field via a smartphone or tablet, and it can also get back to the EOC or ICP using map services where edits can be made and sent back into the system. Enabling the Feature Service capability in the Damage Assessment template reviewed earlier enables this editing of shared data on smartphones and tablet devices, far extending the reach of actionable information in time of crisis.

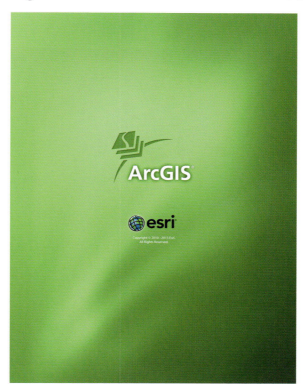

Esri.

A hazards map application viewed on an iPhone using ArcGIS for iOS Esri.

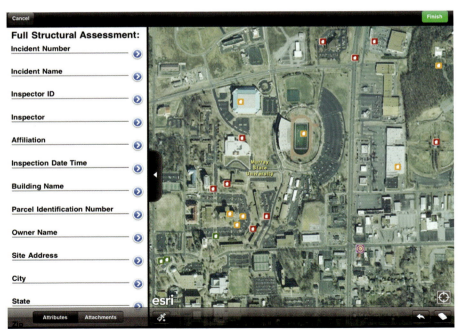

A full structural assessment application administered on an iPad using ArcGIS for iOS

Data courtesy of US Geological Survey.

Application Builder

The new release of the ArcGIS Viewer for Flex 2.5 introduced a new capability, the Application Builder.[19] This tool allows you to configure most of the elements in the COP, such as those reviewed in this chapter, through a wizard-based interface instead of directly editing the respective configuration files. For instructional purposes in the guide, however, we chose to look at the specific XML tags and attributes of each feature, because it gives the user a greater understanding of the details driving the components of each template.

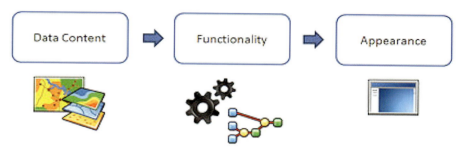

Application Builder workflow Esri.

Conclusion

As you have learned throughout this book, true comprehensive situational awareness during crisis events is enabled by all of the emergency management patterns working together. The LGIM data model provided in the templates is intended as a guideline for a well-thought-out, and complete, *data management strategy.* Achieving this goal lays the foundation for the successful integration and operation of all of the other component parts of the ArcGIS common operating platform. The *planning and analysis pattern* anchors and frames the program. Using sound GIS analytic methods in hazards and vulnerability modeling ensures an effective mitigation strategy in the planning phase, and successful response and recovery operations during and after an event takes place. *Field operations* are effectively coordinated and carried out from a variety of mobile devices that all sync with operations at the EOS and the ICP. All of these emergency management patterns come together to provide a level of situational awareness that could never be achieved without the cooperation of event partners and the sophisticated interoperability of systems and technology. The

---

19 "ArcGIS Resource Center," http://help.arcgis.com/en/webapps/flexviewer/help/index.html#/
Getting_started_with_the_application_builder/01m30000002v000000/. Detailed step-by-step
instructions on how to use the ArcGIS Viewer for Flex application builder can also be found at
http://www.esri.com/news/arcwatch/0212/creating-an-arcgis-viewer-for-flex-step-by-step.html.

COP and its mission-specific map views provide the pathway into and out of the common operating platform that enables this heightened awareness as emergency personnel save lives and protect property in the face of disaster.

A final testimony of survival, post-tornado Tuscaloosa, Alabama, 2011 Courtesy of Karyn Tareen, Geocove.

We continuously update and expand the content around ArcGIS for Emergency Management. To stay current on new information and releases, please follow us on Twitter (@GISPublicSafety), follow our blog[20] on the ArcGIS Resources Local Government—Public Safety Community site, and check the book website for content updates.[21] Please feel free to share your GIS insights and experiences in emergency management via email, blog, Twitter, or Facebook. We also host numerous sessions each year at the annual Esri National Security Summit and Esri International User Conference. There is also the occasional Esri webinar presenting the use and integration of GIS tools in support of emergency management operations and planning. Subscribing to the mailing list ensures notification of upcoming webinars and online training opportunities.

We hope the information presented in this book inspires and supports you in your efforts to prepare and protect your communities from crisis events.

20 http://blogs.esri.com/Dev/blogs/publicsafety/default.aspx.
21 http://esripress.esri.com/bookresources.

Appendix A: ArcGIS for Emergency Management: An Esri white paper

Introduction

ArcGIS for Emergency Management is an openly available baseline configuration of mission-specific templates, tools, and applications sitting on top of the flexible and scalable ArcGIS platform reference architecture. The platform organizes, manages, and delivers appropriate information and data to emergency management personnel based on their specific missions and roles within the organization. As an organization begins to manage its data, perform hazard analysis and risk assessment for its areas of interest, and deliver information in an effective manner, it can begin to respond more effectively and recover more quickly.

ArcGIS for Emergency Management supports and enables common workflows across all aspects of the emergency management mission, from planning to response and recovery. It provides the analytic engine that creates the foundation of good preparedness for an organization by allowing it to conduct comprehensive risk and hazard analysis that identify community vulnerability and highlight mitigation priorities. It also allows better preplanning around events and scenarios that in turn leads to a higher level of overall preparedness. Implementing ArcGIS for Emergency Management promotes enhanced situational awareness to support better decision making by delivering information in a meaningful way for users. This includes alignment with the National Incident Management System (NIMS)/Incident Command System (ICS) and the National Response Framework (NRF).

ArcGIS for Emergency Management provides organizations with a baseline configuration in support of these workflows through the use of templates. These include the following:

- A data model that includes public safety–specific features for operational data
- A common analytic tool and model for vulnerability analysis and impact assessments
- A situational awareness viewer that supports mission-specific delivery of data and tools
- Configuration guidelines for common authoritative data sources such as the Homeland Security Infrastructure Program (HSIP) and HAZUS
- Mobile projects to support emergency management missions, including damage assessment
- A public information map that integrates social media
- A data exchange and catalog portal for collaboration and data discovery

Following is a general description of ArcGIS for Emergency Management and how the ArcGIS platform can be configured to provide the benefits previously described.

ArcGIS for Emergency Management configuration

ArcGIS for Emergency Management is designed to organize and deliver the baseline tools and data typically needed to support an emergency management organization. This delivery includes desktop, server, mobile, and web access using a common repository of data and tools that is managed by the ArcGIS platform. These data and tools have been typically accessed by a single web-based common operating picture (COP). The user experience and access when and where appropriate have been difficult. ArcGIS for Emergency Management promotes access to these data and tools via multiple mission-specific maps or applications that are based on a user's role or responsibilities. These maps and applications are available throughout the organization on any device and are intended to provide targeted and meaningful delivery in support of specific mission requirements. Aligning delivery to a user's mission facilitates the user experience and removes the burden of searching for the right data or tool to answer a specific question. Just as a good traditional paper map helps sift away the noise and focus attention on the task at hand, mission-specific intelligent maps and applications can do the same thing in a digital environment, from any location in the field to the office. These mission-specific views can then be used on both a daily basis and during incident support. The following graphic outlines

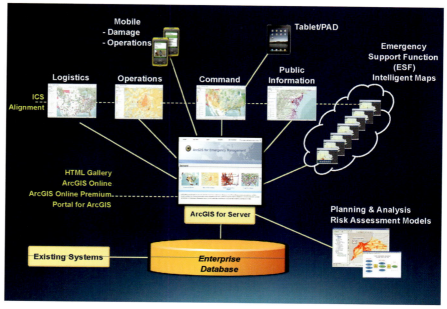

a baseline configuration of the ArcGIS platform—the common operating platform—that aligns with a common emergency management organization structure and mission.

Data management

Emergency management relies on a multitude of data that drives analysis and helps inform decision makers. Creating, editing, and managing this data is the foundation of ArcGIS as the common operating platform. Data comes from many different sources (spreadsheets, web services, business systems, etc.) and can quickly overwhelm decision makers. However, normalizing the data by geography presents patterns that become apparent—tabular data begins to tell a story that supports decision makers. Once data is captured, personnel can begin to feed analytic models and drive intelligent maps or applications that provide insight into how an event is unfolding or a decision impacts an organization. But data management is not just about consumption. Sharing relevant and authoritative data and information products with organizational partners is vital, and ArcGIS provides the mechanism to do that. ArcGIS for Emergency Management is built on sound data management and data sharing capabilities that allow users to form common communities that open the door for collaboration.

Planning and analysis

Raw data might not be useful, but analyzed information is. ArcGIS provides the analytic engine to turn the raw data stream into actionable information. Planning and analysis are most commonly performed through the use of ArcGIS Desktop, but the common operating platform allows access to analytics that can be executed from any location. Analysis is critical to supporting the emergency management life cycle. It is how an organization can analyze risk, understand vulnerability, identify mitigation priorities, develop comprehensive response plans, and test the impact of different event scenarios. The following are examples of common planning and analysis functions for emergency management:

- Conducting a jurisdictional vulnerability analysis based on the occurrence or presence of several elements:
    - Critical infrastructure
    - Natural hazards
    - Technological hazards
    - Historical risk
    - Vulnerable populations
- Managing resources and preplanning
- Modeling loss estimates and impact analysis for events using HAZUS
- Building incident action plan (IAP) maps
- Planning for special events and promote common incident command and control

Situational awareness and response

As an organization begins to fuse and manage all its disparate data within the common operating platform, delivery and access for situational awareness have historically been achieved through a web-based viewer frequently referred to as the common operating picture (COP). As access to data and dynamic information has increased over the years, the traditional COP has become cluttered with too many datasets and tools so use is difficult. Regardless of what users' roles in the organization are or the workflow they are trying to accomplish, they all use the same COP and are forced to find the data or tool they need to answer a question and move on.

ArcGIS for Emergency Management is designed to change the way content is delivered and allow information to flow across the organization in a direct and concise manner. ArcGIS for Emergency Management addresses this by providing mission-specific maps and applications that tailor situational awareness to the role each decision maker or staff is responsible for. This ArcGIS for Emergency Management configuration, with multiple mission-specific viewers in place of one COP, also aligns with NIMS and delivers data and tools based on the ICS framework. A logistics chief has a very different need and use case than an incident commander does. By aligning each view to the organization structure, each person enters into a user-friendly experience that makes sense based on mission requirements. This can be further tailored to even more specific delivery and aligned to the NRF Emergency Support Functions (ESFs), which are responsible for an even finer-grained detail of the organization. ESF 1—Transportation organizations should only be presented with the data (transportation infrastructure) and tools they need, not a catchall system that appears foreign. The following baseline configuration guidance for ICS command, operations, logistics, and public information and for ESFs 1–15 is provided as a starting point for better situational awareness. The high-level design of these views is as follows:

Command executive dashboard

The command dashboard provides situational awareness for decision makers and command staff in a dashboard-type environment. This view is always on and running as a high-level overview of a jurisdiction and the current status it faces. It includes major hazard feeds (weather, earthquake, tsunami, hurricane, etc.) but can also be connected to crisis information management system (CIMS) or computer-aided dispatch (CAD) data to show high-impact events in the community. As a reference to assess the impact of these events, the view should contain the risk/hazard analysis data layers for the organization and the vulnerability analysis conducted. This is not a heavy-lifting viewer with a lot of data or tools but a clean interface designed for maximum consumption or situational awareness without the noise. It should be focused solely on command, departmental executives, or elected officials.

Operations/tactical planning view

The operations/tactical planning view provides a picture of the ongoing situation and response within a jurisdiction. It is the heaviest viewer for tools and data and would contain the most information about the operational aspects of an incident. It should provide tools that allow operations staff to manage field operations, answer questions about impact, illustrate and convey planned activities, and monitor response efforts such as search and rescue or damage assessment. This view should be connected directly to the field personnel who are collecting information in real time. It should also report the status of activities in relation to a stated goal. As the operations staff makes decisions, the ability to mark up the map with incident symbology that conveys status should also be included.

Logistics/resource view

Logistics officers need visibility into the status, availability, and location of resources while working on the management of resources for an organization. This workflow includes the ability to query and task relevant commercial or emergency rental resources potentially needed for events (dump trucks, potable water, portable toilets, lumber, etc.) based on location in proximity to an incident location. The logistics viewer also provides updated information on the transportation network for appropriate routing and movement of resources in addition to incident information and needs coming from the operations and planning sections.

Public information view

During crises, communication with the public is critical to both educate and inform on current plans, activities, and decisions. In return, the public can provide valuable information to the organization and, in essence, become a force extender on the ground. The public information view provides awareness to citizens regarding response and recovery aspects. This viewer is targeted to non-GIS users and is intended to be lightweight, with limited tools and data for the public to access. The data provided will be a subset of the larger operational data that has been approved for public release and disseminate value-added alerts or operational data to inform. Likewise, the viewer can be adapted to collect information from the public through controlled entry forms for volunteered geographic information (VGI) from the crowd or to harvest public information from social media sites (e.g., Twitter, YouTube, Flickr).

ESF viewers

When emergencies occur, the Federal Emergency Management Agency (FEMA) and many states organize around 15 ESFs as defined in NIMS. These represent functions and services that are critical for incident management and recovery. ESFs also become the communication channels between all levels of governments for specific tasks. Each ESF plays a specific role, and therefore, these viewers provide users with more targeted delivery during an incident. As an example, ESF 1—Transportation is responsible for aviation/airspace management and control, transportation safety, restoration/recovery of transportation infrastructure, movement restrictions, and damage and impact assessment to infrastructure. The ESF 1 mission requires specific data and tools that are much different than those of ESF 9—Search and Rescue, which is responsible for lifesaving assistance and conducting search-and-rescue operations.

Briefing view

When an incident briefing is required, it is often done by various officials representing the ESFs above and commonly uses static presentations based on text only. ArcGIS for Emergency Management aligns the data to the ESF mission and, as a result, creates dynamic maps in support of briefings and decision support. Using a tool like ArcGIS Explorer Online in presentation mode, an organization can show the different ESF views dynamically, along with the high-level ICS views, to support live briefings during an operation. The data is current, and the content is relevant without having to stop operations to build briefing materials. All these maps are being updated in real time as users conduct their missions.

Mobile

A final component of ArcGIS for Emergency Management, and perhaps the most critical, is the mobile component. Building tools and applications that work in the field and empower knowledge workers to complete their workflows in a more streamlined manner is an important part of completing the information life cycle. These mobile applications connect the field to the office using the same common operating platform and are largely deployed in support of command and control (incident management), response (search and rescue; situational awareness), and recovery (damage assessment; debris removal) workflows. As workers in the field access the application on their mobile devices, they see the same symbology and data that those in the office are using. As they begin their work, the application should align to their mission by providing only the data and collection tools needed, thereby removing any irrelevant information.

Summary

ArcGIS for Emergency Management is a scalable reference configuration of a common operating platform with viewers and tools designed to support general emergency management workflow requirements. ArcGIS for Emergency Management is designed to deliver content and allow information to flow across an organization in a targeted and meaningful way. By deploying the ArcGIS platform and developing the views and tools described in this document, an organization can truly achieve visibility into all facets of the emergency management life cycle. As organizations begin to shift focus on deploying a common operating platform, they can start to engage users in a meaningful and mission-specific way. This means aligning with common standards (NIMS/ICS) and workflows in emergency management and delivering mission-specific maps and applications directly to the knowledge worker without the need for GIS training. Moving the focus from a picture to a common operating platform not only enhances situational awareness but also empowers users by providing better understanding, collaboration, visualization, and rapid dissemination of critical information when and where it is needed most.

Appendix B: Emergency service function features

Feature Dataset	Feature Class	Command	Operations	Plans	Logisitics	Public Information	ESF 1 - Transportation	ESF 2 - Communications	ESF 3 - Public Works and Engineering
EmergencyOperations	AccessPoint		X	X			X		X
EmergencyOperations	AssignmentBreak	X	X	X					X
EmergencyOperations	AssignmentBreakLabel	X	X	X					X
EmergencyOperations	Division	X	X	X					X
EmergencyOperations	Branch	X	X	X					X
EmergencyOperations	DamageAssessment		X	X					
EmergencyOperations	EvacuationArea	X	X	X			X		
EmergencyOperations	OpsIncidentArea	X	X	X	X	X	X	X	X
EmergencyOperations	OpsIncidentLine	X	X	X	X	X	X	X	X
EmergencyOperations	OpsIncidentPoint	X	X	X	X	X	X	X	X
EmergencyOperations	HelicopterLZandDP		X	X					
EmergencyOperations	Plume	X	X	X					
EmergencyOperations	ResourceAssignment								
EmergencyOperations	RoadBlock	X	X	X	X		X		
EmergencyOperations	USNationalGrid		X						
PublicSafetyPlanning	EmergencyFacility	X	X		X				
PublicSafetyPlanning	HistoricDamageAssessment								
PublicSafetyPlanning	PublicSafetyResource		X		X				X
PublicSafetyPlanning	SpecialEvent	X	X	X	X	X	X	X	X
PublicSafetyPlanning	SpecialEventArea	X	X	X	X	X	X	X	X
PublicSafetyPlanning	SpecialEventLine	X	X	X	X	X	X	X	X
PublicSafetyPlanning	SpecialEventPoint	X	X	X	X	X	X	X	X
PublicSafetyPlanning	VideoFeed				X		X		
CitizenService	ServiceRequest				X				
FieldCrew	AGM_FieldCrewMembers		X						
ReferenceData	FacilitySitePoint						X	X	X
ParcelPublishing	OwnerParcel								
	Post-Event Imagery	X	X			X			

ESF 4 - Firefighting	ESF 5 - Emergency Management	ESF 6 - Mass Care, Emergency Assistance, Housing, and Human Services	ESF 7 - Logistics Management and Resource Support	ESF 8 - Public Health and Medical Services	ESF 9 - Search and Rescue	ESF 10 - Oil and Hazardous Materials Response	ESF 11 - Agriculture and Natural Resources	ESF 12 - Energy	ESF 13 - Public Safety and Security	ESF 14 - Long-Term Community Recovery	ESF 15 - External Affairs
	X		X		X				X		
					X						
					X						
					X						
					X						
	X										
	X				X				X		
X	X	X	X	X	X	X	X	X	X	X	X
X	X	X	X	X	X	X	X	X	X	X	X
X	X	X	X	X	X	X	X	X	X	X	X
X					X				X		
	X										
	X		X						X		
					X						
	X	X		X							
	X		X						X		
X	X	X	X	X	X	X	X	X	X	X	X
X	X	X	X	X	X	X	X	X	X	X	X
X	X	X	X	X	X	X	X	X	X	X	X
X	X	X	X	X	X	X	X	X	X	X	X
		X	X								
											X
				X							
X	X	X	X	X	X	X	X	X	X	X	X
				X							
	X	X		X							X

Appendix C: Emergency service function widgets and tools

Viewer/Category	Tool	Command	Operations	Plans	Logisitics	Public Information	ESF 1 - Transportation	ESF 2 - Communications	ESF 3 - Public Works and Engineering
Public Safety Resource Center COP	ERG						X		X
	National Grid								
	Find Closest Resource				X		X	X	X
	Report by Exception								
	Critical Infrastructure								
	Bomb Threat								
	PDC Active Hazards								
	Config file switcher								
	Event Picker								
	Imagery Swipe								
	Routing						X	X	X
ArcGIS Viewer for Flex Core	Edit						X	X	X
	Chart				X			X	
	Print								
	Query								
	Search								
	Data Extract								
	Draw								
	GeoRSS								
Community	Social Media								
	WMS Radar								
	Elevation Profile								

ESF 4 - Firefighting	ESF 5 - Emergency Management	ESF 6 - Mass Care, Emergency Assistance, Housing, and Human Services	ESF 7 - Logistics Management and Resource Support	ESF 8 - Public Health and Medical Services	ESF 9 - Search and Rescue	ESF 10 - Oil and Hazardous Materials Response	ESF 11 - Agriculture and Natural Resources	ESF 12 - Energy	ESF 13 - Public Safety and Security	ESF 14 - Long-Term Community Recovery	ESF 15 - External Affairs
X	X					X					
	X				X						
X		X	X		X					X	
	X					X					
	X	X	X	X		X	X	X			
	X										
					X						
X		X	X	X	X				X	X	
X	X	X	X	X	X						
	X										
	X										
	X										X
	X										